Inventors
Project Book

BY

L. GEORGE LAWRENCE

HOWARD W. SAMS & CO., INC.
THE BOBBS-MERRILL CO., INC.
INDIANAPOLIS · KANSAS CITY · NEW YORK

FIRST EDITION

FIRST PRINTING—1971

Copyright © 1971 by Howard W. Sams & Co., Inc., Indianapolis, Indiana 46206. Printed in the United States of America.

International Standard Book Number: 0-672-20827-X
Library of Congress Catalog Card Number: 74-157799

Preface

In America today, inventing is one of the most wonderful pastimes available to us. Aside from projecting a continuous and fascinating challenge throughout life, it bears a most viable seed of the million-dollar American dream. Fortunately, you do not have to be a "genius" to make this dream come true! Genius, according to the late Edison, is little more than "1 percent inspiration and 99 percent perspiration." Thus, while certain inductive and deductive processes are involved, invention and its rewards can be enjoyed by all.

The current volume was triggered by the wide acceptance our earlier *Inventors Idea Book,* now out of print, received. We are presenting new material in this book, but have maintained our previous manner of presentation, giving problems, design outlines, methods, references where applicable, and tentative solutions to mark points from which to start.

The bulk of the material is new and has never been published before. Owing to queries and requests by readers, we have added new sections on background research and resources. Other entries pertain to such important and diversified topics as selling of inventions, invention brokers and agents, foreign patent offices, invention fairs, protection of priorities, government contracts, research and military specifications, rental of laboratory equipment, copyrights, trademarks, inventions made by military personnel, and new discoveries pertaining to stunning natural phenomena known only to a selected few.

In presenting the *Inventors Project Book,* we hope that it will function as a dynamic guide and money-making tool for both novice and seasoned inventors. It is specifically directed at those creative men and women who not only desire to contribute something new, but are willing to carry projects to successful completion.

L. GEORGE LAWRENCE

Imagination is more important than knowledge.—Albert Einstein
(1879-1955)

Contents

APPENDIX A

APPENDIX B

APPENDIX C

APPENDIX D

APPENDIX E

APPENDIX F

Invention at Large

Total dependence on invention is a pre-eminent characteristic of our civilization. Invention is the key to successful living. Take the inventions away, and mankind would revert to stone-age barbarism overnight.

As with everything else our race has created, the chief purpose of invention is to gratify both general and selected human needs and desires for true reasons yet to be defined. Ancient legends and religions imply an extraterrestrial origin of invention, principally designed to divorce early man from his simian ancestry. Here we encounter tremendously wise and dynamic beings like the amphibious god *Oannes,* who gave basic knowledge to the Sumerians in the year 4000 B.C. We also encounter benevolent and efficient work of the Egyptian goddess *Isis,* best known for inventions in agriculture and the introduction of wheat and barley to inhabitants of the Nile regions. The red-bearded *Quetzalcoatl*—sky-god, inventor, and legendary ruler of Toltec in Mexico—added his prolific work to an exciting and vast inventory of mythological inferences. Like the other "gods," he too disappeared as suddenly as he came.

However, taken together, the messages contained in what might appear to be some odd and special missionary work performed by some powerful galactic federation, have retained a dominant validity to this day. For these ancient stories and legends, regardless of how we evaluate them, embody important guidelines for us to follow. Specifically, we are directed to recognize the distinct characteristics common to successful inventions: pre-eminent benefits to society at large, improvement of survival aspects, generation of employment, and the ability to make economies grow.

Today, invention is probably more important than ever before. We have entered an era in which machine is pitted against machine. Once-fabulous inventions have commenced to pollute our environment in a manner similar to chemical warfare directed against us by foreign enemies. Technological ways and means are needed to attenuate these severities and, simultaneously, to advance both hardware and concepts for more healthful living.

As a result of demands made on innovative manpower, the 1970's will require almost twice as many inventor-type technologists as we had in the 1960's. This need implies that more rewarding opportunities will exist not only for corporate inventor teams, but for the "lone" inventor as well.

By and large, fortunately, the element of "genius" in invention is an exception rather than a rule. Today, many highly successful inventors owe their fame and fortunes not so much to superior skill in solving problems as to wisdom in choosing them. To that end, it is worth your time to consider the particulars and basic approaches on which choices can be based.

DATA RESOURCES

In invention, it is imperative that new ideas be viewed in legal terms. You are not alone in the invention business, and the danger of patent infringement is always near. For these and other reasons, background research suggests itself as a precautionary measure. But it has other benefits as well. It is an excellent means for discovering what other inventors have done, and thus it points out unique avenues of approach and leads to better and more advanced ideas.

During preliminary search phases, many inventors find it economical to do the pertinent tasks themselves rather than to employ a patent attorney. He essentially works with the same background material listed below and elsewhere in this book.

For those who desire aids for an in-depth patent search, the following list of addresses and particulars pertaining to foreign patent offices has been prepared.

Australia: Commissioner of Patents, Patents Office, Canberra, A.C.T.; 50 cents (Australian) per copy of patent, postpaid by surface mail.

Austria: Oesterreichisches Patentamt, Druckschriftenverschleiss, Kohlmarkt 8-10, Vienna 1; 12 Austrian shillings per copy of patent (for photocopies, 4.90 Austrian shillings per page *after* the original edition has been sold out), plus postage for air or surface mail. An account may be opened with an initial deposit of 1000 Austrian shillings.

Belgium: Service de la Propriete Industrielle et Commerciale, 19 rue de la Loi, Brussels 4. Reproductions of patents are available as Xerox copies at 7.5 BF per page or as microcards (1-4 pages per card) at 10 BF per microcard. Foreign postage is charged. Advance deposits may be arranged.

Canada: Commissioner of Patents, Ottawa; $1.00 per copy for patents issued as of January 1, 1948, beginning at No. 445,931. Reproduced patents prior to 1948 can be purchased for $0.25 per page.

Czechoslovakia: Dr. Karl Neumann, Advokatni poradna c. 10, Zitna 25, Prague I; Kcs. 10 for single copies, approximately Kcs. 7.40 for more than one copy.

Denmark: Direktoratet for Patent- og Varemaerkevaesenet, Nyropsgade 45, Copenhagen V; 4.00 Kr. per copy, plus postage. Will accept advance deposits.

England: Comptroller-General, The Patent Office (Sale Branch), Block C, Station Square House, St. Mary Cray, Orpington, Kent; 4S. 6d. per copy (photocopies of out-of-print patents are available for the same price). Payment by International Money Order. Advance deposits of £4 or more accepted.

Finland: Patentii- ja Rekisterihallitus, Bulevardi 21, Helsinki 18; fmk. 1 per copy, plus postage.

France: Service d'Edition et de Vente des Publications Officielles, 39 rue de la Convention, Paris 15; 2 F per copy, plus 0.7 F postage for 1 copy. Advance deposits of 200 F or more accepted if orders are placed frequently.

Germany (East): Zeitungsvertrieb Gebrueder Petermann, Roennebergerstr. 1, Berlin 41; DM 1.50 (West German Marks) per copy.

Germany: (West): Deutsches Patentamt, Dienststelle Berlin, Gitschiner Str. 97-103, 1000 Berlin 61; DM 2.00 per copy, plus postage, for either granted patents or published patent applications. Advance deposit of DM 30 or more accepted if orders are placed frequently.

Hungary: Licencia Hungarian Company for the Commercial Exploitation of Inventions, P.O. Box 207, Budapest 5. (No fees stated.)

India: Patents of number 1-75,000: Controller of Patents and Designs, 214 Lower Circular Road, Calcutta 17. Patents of numbers 75,001 and higher: Officer-in-Charge, Government of India Book Depot, 8 Hastings St., Calcutta 1. Printed specifications published prior to November 1, 1950, Re. 1 per copy; those published on or after that date, Rs. 2 per copy. Foreign postage is charged.

Israel: The Registrar General, Registrar of Patents and Designs, P.O. Box 767, Jerusalem; 70 agorot per page, plus postage.

Italy: Libreria dello Stato, Piazza G. Verdi, 10 Rome; L. 200 per copy for 8 pages or fewer, L. 200 for additional 8 pages or fewer plus postage; photocopies of out-of-print patent specifications, L. 200 per page. Advance deposits accepted.

Japan: Hatsumei-Kyokai (Invention Association), c/o Patent Office, 1, San-nen-cho, Chiyoda-ku, Tokyo. Rates for foreign clients: individual copy, $0.05 per page; photocopy, $0.20 per page, plus postage. Most of the older specifications are available only as photocopies. Advance deposits accepted. For information on patent searches, translations, and publications, write to Patent Information World-Wide Service, Invention Association, 17 Shiba Nishakubo Akefune-cho, Minato-ku, Tokyo.

Netherlands: The Patent Office (Octrooiraad), Willem Witsenplein 6, The Hague; patents, f 1.50 per copy; patent applications laid open to public inspection, f 2.50 per copy, postpaid by surface mail. Coupons for use in purchase of specifications are available in minimum amounts of f 75.00 and f 125.00, respectively. Advance deposit of f 300.00 accepted.

Norway: Styret for det industrielle rettsvern, P.O. Box 5085 Majorstua, Oslo 3; Kr. 1.50 per copy plus postage. Advance deposits of Kr. 100.00 or more accepted.

Poland: Urzad Patentowy Polskiej Rzeczypospolitej Ludowej, Warsaw 68, al. Niepodleglosei 188; $0.50 per copy, including cost of handling and postage by registered mail.

Spain: Photocopies of patents are available from Spanish patent agents who are members of the Colegio Oficial de Agentes de la Propiedad Industrial. Cost is $3.00 for the first 4 sheets, plus $0.60 for each additional sheet. Requests for names and addresses of these agents should be directed to Colegio Oficial de Agentes de la Propiedad Industrial, Montera 13, Madrid 14.

Sweden: Kungl. Patent- och Registeringsverket, Stockholm 5; Kr. 3:00 per copy of patents that contain drawings, otherwise Kr. 2:50. Mailed postpaid.

Switzerland: Eidgenoessisches Amt fuer geistiges Eigentum, 3003 Berne; Fr. 1.50 per copy plus postage. Coupons for use in purchase of specifications are available in pads of 10 at Fr. 1.40 or in pads of 100 at Fr. 1.30 per copy (air mail postage included).

United States: Commissioner of Patents, Washington, D.C. 20231; $0.50 per copy, payable in advance. Coupons for use in purchase of specifications: $0.50 coupons sold in pads of 10 for $5.00, 50 with stub (for record) for $25.00.

Photocopies of many foreign patents may be obtained from the U.S. Patent office at a cost of $0.30 per page. Subjects of interest may by identified in patent gazettes, stating name of invention and inventor, number of patent, and date of issue.

If your invention is highly science-oriented and will ultimately be tailored for governmental use, the very latest of research reports might be necessary for your work. Reports of this type normally cannot be found in the holdings of university libraries, nor can details of a given specialty be learned from the *Official Gazette* published weekly by the U.S. Patent Office. For this purpose, you can use the services of the Science Information Exchange (SIE). Its full address is:

Science Information Exchange (SIE)
300 Madison National Bank Building
1730 M Street, N.W.
Washington, D.C. 20036

Established in 1949, SIE has operated since 1953 under the aegis of the famous Smithsonian Institution. Fig. 1-1 shows SIE's overall role as a national information network. Over 1000 research organizations, including federal agencies, private foundations, universities, state and city governments, and foreign sources cooperate and/or participate. SIE operates on a fee basis. Coverage includes all fields of basic and applied research, as well as the biological, medical, agricultural, physical, engineering, and social sciences. Clients fill out a "Request for Services" form furnished to them, stating the specific research or problem on which information is desired.

Fig. 1-1. National information network of Science Information Exchange.

A number of commercial patent-search agencies, usually located in Washington, D.C., have emerged. At a cost of about $6 per search, this is an inexpensive service that can be used to your advantage. Normally, a minimum of three copies of patents close to your idea will be sent postpaid. These agencies also offer preparation of patent applications, with prices ranging from $240 to $360. However, if search services are provided by a patent attorney's staff, you will receive about seven copies. Fees range from $40 to $110 or more. Most inventors elect to have their final applications for patent prepared by patent attorneys. In the case of a local attorney, for example, it is possible to inspect the work in progress, and perhaps make last-minute changes as the specifications and claims are spelled out and drawings are made.

Emphasis has been placed on the preliminary search because it can save time and trouble later. Also, in order to invent gainfully, it is necessary to know the state of the art in depth. The degree of intensity of background searches is, of course, determined by the character of your invention. For example, if yours is a discovery-type invention involving new principles of physics and the like, background investigations will revolve around this new principle, and frequently are less costly. But if you deal with a known, established art, you might have to sweep for novel references internationally.

ANALOGIES AND HUMAN NEED

With respect to inventions, mere ideas are worthless until they have been proved practical by application. It is also unfortunate that untrained inventors rush to conclusions too quickly. The pride of having discovered, or seemingly discovered, something new is a pride that often conquers all logic, commercial sense, and reason. Or there might be interferences by the inventor's family, bias by friends, or false advice by shyster invention brokers.

Top-selling inventions are not necessarily *totally new* inventions but exceptionally good improvements of an existing object or technology. Totally new devices or methods require much groundwork, typically in the form of promotional publicity, advertising, or a good number of pilot sales. The world is changing, and new needs and requirements arise daily. But if the inventor or manufacturer ventures to accelerate this change beyond a certain point, he is bound to meet strong resistance.

Therefore, for good reasons, it is necessary to examine the world of technological examples and analogies. The dictionary defines the latter as "(1) similarity in some respects, (2) a comparing of something point by point with something else, (3) in *logic:* the inference that certain resemblances imply probable further similarity."

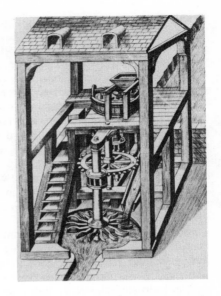

Fig. 1-2. Medieval water mill with Pelton-type mill runner (bottom).

While the above is no "secret recipe" for successful invention, it permits an unburdened mind, fresh and clean and equipped with the bare fundamentals of orthodox physics, to find the path of least resistance between an idea and the condensation of that idea into practicality.

As we shall see by means of analogies and guiding examples given below, some of our most fascinating and useful modern-day "inventions" are revived *composites;* that is, a number of seemingly

Fig. 1-3. Modern Pelton runner.

Courtesy Baldwin-Lima-Hamilton Corp.

dated concepts of classical origin have been re-evaluated and refined to bring modern devices into being.

In the area of information-display technology, for example, no really great strides could be made because of the lack of electronics. Other machine systems required advanced mathematical tools and research instruments to give them a new, more efficient status.

This dramatic evolutionary process, the problem-detecting and solving capacity of inventors at large, has enjoyed a long and continuous history. We have mentioned some really basic inventors like Oannes and Isis. However, in the realm of machine-oriented innovations, excellent prototype analogies emerged in the 15th century. Jaques Besson, a French inventor, was the first to succeed in publishing his machine concepts. Vittorio Zonca followed; his works were published posthumously by the Paduan printer Francesco Bertelli in 1607. Zonca enchanted his age with some of the most fabulous (though nonfunctional) perpetuum mobiles ever conceived. However, while apparently useless, they involved marvelous gear systems and principles of leverage. So, it is out of these and other legacies that designs of modern technology were destined to evolve.

The following classical renditions are good guides for our purpose, showing how seemingly forgotten concepts can be used successfully. They have in common the gratification of both general and selected human needs and desires—which are typical ingredients of successful inventions!

Fig. 1-2, for example, shows a water mill which gave tremendous support to the economies of scattered villages in medieval Europe. The efficiency was low, so carpenters in the Alpine and Pyrenean valleys of southern France continued to experiment with paddles of the impulse-type mill runner (bottom) in an effort to use every drop of water. This development, which was entirely empirical and based upon trial-and-error methods, finally culminated in the design of a superior machine: the Pelton runner shown in Fig. 1-3. Mounted on a *horizontal* shaft, this design now can develop 18,500 horsepower at only 225 r.p.m. Major improvements in efficiency were made by changing the geometry of the buckets and using a nozzle-type water injection system. Both simplicity and reliable, long-term performance were the outcome of an old and very dated original concept.

Fig. 1-4 shows the process of manufacturing paper in medieval times. The final products of this art were restricted in overall use and application. Originally, almost up to modern days, paper was little more than basic material for printers, writers, packers, and decorative trades. Early electricians gradually discovered its insulating properties. Today, paper is used in some 14,000 products. And water, heretofore considered an arch-enemy of paper, is conquered by the paper towel. It is somewhat striking that the towel-like drying

Fig. 1-4. Medieval paper mill used manual press, towel-like sheet drying process.

process of paper sheets, as seen in Fig. 1-4, did not suggest a converse use at that time.

High-speed manufacture of paper was afforded in 1798 by machines forming continuous webs. These replaced the cumbersome dipping mold shown in Fig. 1-4.

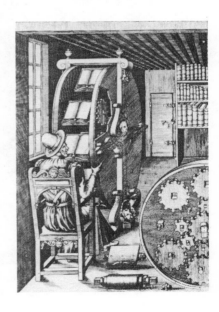

Fig. 1-5. Ramelli's reading machine featured basic idea of high-speed data access.

Another good example of recognizing and solving a problem is afforded by educational machines. In eras past, education was a privilege of upper classes only. The children of day laborers and paraprofessional tradesmen barely progressed beyond the knowledge to write their names, count a series of numbers, and recognize the value of money and goods. Ramelli's "mechanical reading desk," shown in Fig. 1-5, was invented in the 15th century and was intended to present bulk information to well-to-do families and their children. Though efforts were made to expand its use, the desires of Renaissance educators went remarkably unfulfilled. It took modern electronics and audio-visual teaching technology to transform Ramelli's basic concept into a functional data retrieval system. Today, as illustrated in Fig. 1-6, a young child can be exposed to self-operated teaching systems using prepared input materials. However, as in book-type learning methods, note that the principal changes involve the method and access technique used. Ramelli's basic idea, rapid presentation of bulk-data inventories, has been retained.

The above shows how well-known or classical elements have been sifted, evolved, and combined, giving a new expression or synthesis. To that end, if we consider our three analogies on a grand scale, our efforts in invention are guided by realizing that practically

Courtesy International Business Machines Corp.

Fig. 1-6. Modern audio-visual data retrieval/display system.

every machine, process, or device used today will eventually have to be invented all over again. Indeed, though ours might be an age of wonder, it is far from being an age of technological efficiency. With barely an exception, most of the machines available are strikingly crude, only partially developed, and poor facsimiles of what they really should be.

We have, for example, no nonpolluting thermal power plants, electric aircraft, or heavy-duty automotive engines. We are not blessed with simple concepts pertaining to wheel-less surface vehicles; we have no way of transmitting dc power by wireless means. There are no *organic* computers (outside of the human brain) endowed with innate reasoning capabilities. Nor are there corrosion-proof ships, nonmessy feeding utensils for babies, or means for allowing women to arrange their hairdos without the involved expediency of chemicals and heat. We do not even have a really safe, inexpensive, and steerable baby stroller!

Human progress demands that we use less wasteful articles and machines—less wasteful not only in areas of pollution and energy budgets allowed for them, but of time and applied labor as well.

While it is true that inventions involving the use of natural phenomena have a "classical" character and are difficult to modify, the same cannot be said of man-made techniques. Practically all of these various fields and aspects are wide open for improvement and invite discerning minds.

The inherent complexity of problems changes from one topic to the next, but rare is the invention which cannot be produced, at least as a prototype, in the garage or on a vacant kitchen table.

Invention Sales Program

As the American patent system approaches its 180th anniversary, economists have been studying the rate of inventions quite critically. George Washington signed the first patent law, "An Act to promote the Progress of useful Arts," on April 10, 1790. During the same year, he authenticated the first U.S. patent, involving a machine and process for making potash. Only three patents were granted in 1790. Today, the U.S. Patent Office frequently receives 1000 or more patent applications per week. The number of patents issued now climbs toward 3,500,000 in the United States alone.

From a commercial point of view, an inventor normally takes one of two roads in order to turn his brain child into cash: (1) apply for a patent and sell it for a fixed sum or royalties, or (2) manufacture the article himself. The latter step involves financial resources, since facilities must be acquired, machines must be purchased, personnel must be employed, and so on. It is also necessary to have adequate financial reserves, the latter to be used until returns from the invention produce company income.

Since strong financial backing frequently is not available to the lone inventor, efforts are made to sell the invention to a prospective buyer. An agent or invention broker may be employed to perform the necessary "leg-work"; that is, preparation of sales literature on the basis of outlines furnished by the inventor. In all fairness, it should be pointed out that patented or patent-pending inventions carry a greater sales potential than mere, unprotected ideas. Therefore, if at all possible, the inventor should strive to obtain a formal patent for his invention.

Some general considerations attached to patents are as follows:

Q. Can an invention be protected without a patent?

A. No. The only way to protect an invention is to secure a patent on it, for as long as it is unpatented, even though the inventor is entitled to a patent, anyone may make, use or sell it without the inventor's permission.

Q. Does the inventor have to go to Washington to apply for a patent?

A. No. To start patent proceedings the inventor need only send a disclosure of his invention in the form of a model or sketches and description to the patent engineer. All correspondence pertaining to an application for patent can be conducted through the mail.

Q. Does the inventor have to wait until he obtains his patent in order to sell or produce his invention commercially?

A. No.

Q. What do the terms "Patent Applied For" and "Patent Pending" mean?

A. These terms are commonly used by inventors and manufacturers to serve notice that an application for patent is pending in the U.S. Patent Office and when the patent is issued, infringers may be sued for damages. They have no actual legal force but have a practical effect in warning possible imitators that they may be held liable for infringement in the event a patent is granted.

Q. Does the U.S. patent protect the invention in foreign countries?

A. No. A U.S. patent protects the invention only within the United States and its territories. In order to protect an invention in foreign countries, a separate patent application must be filed in each foreign country in which the inventor wishes to protect his invention. There are many restrictions, compulsory working, fees, and annual taxes on foreign patents in most countries. Information about foreign patents and the propriety of filing should be available from the patent engineer who handles the U.S. patent application.

Q. What happens if the inventor makes changes in his invention after obtaining his patent?

A. If an inventor has made an improvement in his invention after he has obtained a patent, a new application for patent must be filed in the Patent Office provided the improvement is patentable over the original invention. The inventor should not hesitate to make changes that improve his invention. He should inform his patent engineer as soon as he has perfected an improvement so that an additional patent application may be prepared and filed, if advisable.

No one has the right to make a device claimed in an unexpired patent without the permission of the patentee, even though the maker wishes to construct the machine solely for his private use, and not for sale.

If an invention is protected by a patent in this country, it cannot be manufactured in another country and then imported, sold, or used in this country without a license from the patentee.

The life of a patent (other than a design patent) is seventeen years from the date of its issue, and upon its expiration, it cannot be renewed or extended, except by an act of Congress.

It is a well-established principle that an inventor has the right to employ the mechanical skill of others to perfect his invention without forfeiting his right to the invention.

He who merely suggests that an invention be made and furnishes the money to do it is not the inventor, as against the person who conceives the idea and reduces it to practice.

The law does not look with favor upon a party who withholds his invention from the public by negligent postponement of his claim until others have made and introduced the same.

Models or sketches made during experiment or development of an invention should never be destroyed. They should be dated when made and saved. They may prove valuable at some later time in establishing the date of conception of the invention.

The relationship between the patent engineer and the inventor is highly confidential, and no information relating to the invention will be given to anyone, other than the inventor or such person as the inventor may designate in writing.

Specific information about patents may be obtained from the U.S. Patent Office, Washington, D.C. An excellent guide, "General Information Concerning Patents," may be obtained at a cost of 20 cents from:

Superintendent of Documents
U.S. Government Printing Office
Washington, D.C. 20404

How reliable are patent brokers or agents? Many of these firms advertise in popular magazines and promise various types of services.

TO WHOM IT MAY CONCERN

       This is to certify that _____
is the inventor of a _____
illustrated in this document. I hereby submit the
following information to (name of broker) for
examination and promotion if found acceptable.

       Permission is hereby granted to (name of broker)
to disclose this information to manufacturers and/or
potential buyers that have shown interest in negoti-
ating the purchase or license of my invention.

       For an outright sale of this invention, I wish
a minimum of $ _____ (or larger sum of money
if possible) or a down payment of $ _____ (or
larger sum of money if possible) plus a royalty
of _____% on the manufacturers' sale price of
this invention.

       I reserve the right to revoke all rights granted
by this document at any time. In revoking the rights
granted herein, prior to acceptance of any sales
offer negotiated by (name of broker), I am under no
financial obligation to (name of broker).

Date_____   Signature of Inventor_____

Signature of Notary Public_____ Date_____

Fig. 2-1. Sample of invention sales form.

The great majority of these brokers are honest men, for their clients' success is theirs as well. Typical service fees range from $60 to $125. These fees normally are spelled out in the broker's initial correspondence with inventor-clients. A commission of about 10 percent is customary for each invention sold; the initial handling fee is refunded as soon as a binding contract has been signed with a manufacturer. Fig. 2-1 shows a basic agreement.

As a rule of thumb, invention brokers prefer to handle simple inventions directed at mass markets. Samples of preferred inventions are shown in Fig. 2-2. Items like the paper clip (A), hair clip (B), crimped bottle cap (C), and sewing-machine needle (D) belong to the class of best sellers. A novelty, spring-driven spit (E) for campers is appealing, but demands a somewhat specialized market. A complex scientific process (F) normally cannot be handled, since its sale involves many tests, large-scale laboratories, and massive resources which only a rich country but not a manufacturer could afford.

The sales process is the most important aspect attached to invention. In the case of brokers, special files may be used in order to place an article with a given firm. The average broker normally uses the U.S. mail for sending descriptions of unsolicited inventions to a selected group of companies. This process involves little else but copying the inventor's description form. Prospective buyers are

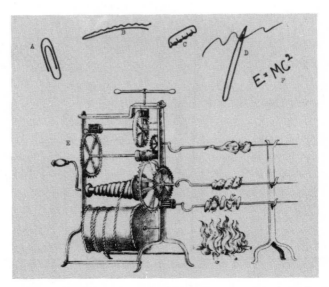

Fig. 2-2. Inventions preferred by patent brokers and sales agents are simple types, but have attributes that make them of best-selling quality.

selected from an index of manufacturers. However, in order to save a good amount of money, many free-lance inventors simply purchase a published index and mail their disclosures themselves. An index that is highly complete is the *Thomas Register of American Manufacturers*. It may be obtained, including upright storage rack, at a cost of $35 from:

Thomas Publishing Company
461 8th Ave.
New York, N.Y. 10001

Most company listings in the *Thomas Register* carry ratings in dollar values. Brokers normally do not provide such ratings.

All inventions made, regardless of their sales outcome, must be safeguarded in ways of priority dates and data. It should be realized that the granting of a patent is nothing more than specific recognition by the U.S. Government of the fact that a certain individual was first to conceive a certain idea and to reduce it to practice. We stress the word *first,* for if an invention is patented, the patent can be rendered invalid if another inventor projects conclusive proof that he was the first one to conceive the idea. A large number of important cases have rested on the proof of priority. As protection against a severe catastrophe without rewards, a general summary sheet should be prepared and, together with certified signatures and dates, kept at a safe place. Entries should include the following items:

1. Name and address of inventor
2. Name of invention or machine
3. Purpose of invention or machine
4. Date conceived
5. Date disclosed:

    (A) Witness (How disclosed and date)
    (B) Witness (How disclosed and date)
    (C) Witness (How disclosed and date)

6. Dates of sketches
7. Dates of written description
8. Dates of working drawings
9. Date of starting on machine
10. Bill of materials and listing of principal vendors
11. Date of completion
12. Dates of photographs, name and address of photographer and/or photofinisher
13. Date tested
14. Where tested; names of witnesses

15. Results of tests
16. Newspaper publicity (if any)
17. Disposition:

    (A) Date matter was placed in hands of patent attorney
    (B) Date matter was placed in hands of broker
    (C) Date invention was offered for sale to selected manufacturers through the mails.

This summary sheet should be augmented by research notes, references, and copies of patents whose subject matter is close to your new invention (copies may be obtained, at a cost of 50 cents, from the U.S. Patent Office). None of these documents should be prepared in a casual, amateurish form. Since professionals might have to pass judgment on your claims in court, the presentation of odd scribbles, incomplete and coffee-stained sketches, etc., is not likely to cast a favorable light.

COPYRIGHT AND TRADEMARKS

Aside from obtaining a patent *per se,* some inventors elect to have their descriptive matter copyrighted as well. A booklet titled *General Information on Copyright,* and application forms for filing for a registered copyright, may be obtained from:

Copyright Office
The Library of Congress
Washington, D.C. 20540

The fee for printed matter, including books, sales bulletins, catalogs, brochures, etc., is $6. Copyright is allowed for print or labels used for an article of merchandise. The U.S. arrangement differs from that in many foreign countries, which frequently do not allow copyrights for items other than works of music or literature.

In regard to invention-type materials, the U.S. copyright would cover your written description, drawings, and photographs, and should be printed or lithographed as your advertisement for your product. If this material is comprehensive enough, it will establish ownership. A registered copyright runs for 28 years in the United States. It may be renewed *prior* to its termination for an additional period of 28 years.

Most manufacturing inventors find it advantageous to create a specific name for their product, resulting in a better public image and repeat orders. Or a *trademark* may be chosen. A booklet titled *Trademarks* may be purchased at a cost of 25 cents from:

Superintendent of Documents
U.S. Government Printing Office
Washington, D.C. 20404

MANUFACTURING

Filling technical and general needs is big business; the larger the need for an item, the greater are its chances for commercial success. Unfortunately, as was pointed out earlier, the lack of financial resources frequently prevents highly talented inventors from turning their ideas into cash. The following considerations might be useful.

Assuming that basic protection of the invention has been taken care of, prototypes of the invention can be fabricated on the proverbial kitchen table and offered to distributors of respective product lines. Thus, if an appreciable number of written orders are received, a low-interest bank loan may be obtained against this collateral. The status of the collateral will be improved once machines have been purchased for manufacturing purposes and formal business activities have been established. In time, the manufacturing inventor delegates sales operations to sales personnel of his own, or, for the sake of economy, he has salesmen add his product to their traditional lines.

However, it is important to avoid, if at all possible, a one-product line because it may not survive recurring business cycles. If a number of different articles, for different uses, can be made on the same kind of shop equipment and with the same type of raw materials, long-term sales and business stability will be enhanced. Injection-molding techniques, for example, allow excellent diversification, with final products ranging from household goods to control knobs for electronic apparatus.

Highly specialized inventions, typically machines having a one-of-a-kind character, can be constructed and rented out to seasonal users for a fixed fee and/or royalties computed on a piece basis. An electronic strawberry sorter is a seasonal machine. So is a special dredging system for cleaning drainage pipes electrically during a low-flood period of a river. Note, however, that invention brokers normally do not handle rental operations. Typically, the existence and availability of a specialized machine are made known to prospective users by personal contact and in-plant demonstrations, or through advertising in the trade press.

UNSOLICITED PROPOSALS AND RESEARCH

Many independent inventors, particularly electronics people, find it rewarding to evolve special machines and equipment for government use, typically for defense and peripheral purposes. A manufacturing program of this kind entails obtaining a federal security clearance in order to be permitted to handle classified material. Here, the inventor's background is investigated in depth on the basis of references furnished by the clearance applicant. A "secret" or other clearance classification must be obtained before manufacturing operations can begin.

Government business is afforded by means of bid lists or gazettes provided by federal agencies. A typical list contains descriptive abstracts of items needed, and the name of the agency soliciting bids. The most reasonable bid is accepted. However, military procurement is subordinated to extremely demanding quality standards, the MIL SPEC.

Typical federal supply classifications are:

Group 58: Communications Equipment
Group 59: Electrical and Electronic Equipment Components
Group 61: Electric Wire and Power and Distribution Equipment
Group 66: Instruments and Laboratory Equipment.

All component parts used in manufacture, as well as the final unit or machine itself, must meet specific performance and test requirements. An index of *Military Specifications and Sources* can be obtained from:

Military Specifications & Sources
801 North La Brea Ave.
Los Angeles, Calif. 90038

The index is published quarterly in January, April, July, and October, with specific termination dates listed. The cost is $25 per year ($30 per year outside the United States).

While most government-connected manufacturing operations are traditional ones, inventor-owned shops can do well by projecting an *unsolicited* bid to a given agency. The proposal incorporated in such a bid might pertain to a new, highly advanced item suited for defense, effective environmental control, and the like. If the invention is needed and lives up to its promise, the possibility of purchase plus award of a research grant is excellent.

Basically, in order to prepare an unsolicited proposal, a given topic must be researched in depth and supported with experimental data, and references must be cited. Furthermore, the proposal must include a clear delineation of factory capabilities, brief biographies of principal engineers, and a detailed estimate of cost. The latter item may be computed on a "cost-plus" basis, i.e., actual cost of the project plus 3 percent (or more) profit margin. If complex but exceptionally functional systems are involved, good profits can be realized. Also, since the government tends to furnish special equipment items free of charge, it is possible to enlarge a laboratory instrumentation pool. However, diligent background research is required in order to lend convincing substance to unsolicited proposals.

Established in 1962 in the Library of Congress with support of the National Science Foundation, the *National Referral Center for Science and Technology* was given the responsibility of identifying all significant information resources in the fields of science and technology. This task included acquiring data describing the specialized capabilities of these resources, and providing guidance about their use. As a result, the *Directory of Information Resources in the United States* came into being. Copies may be obtained, at a cost of $2.25, from:

Superintendent of Documents
U.S. Government Printing Office
Washington, D.C. 20404

The entries list libraries and other information centers of both government agencies and large corporations in alphabetical order. Restrictions on circulation of materials, loan privileges, and the like are included.

Information may be augmented by technical literature available at local university and college libraries catering to one or more schools of engineering. Some selection is required, since liberal-arts colleges, if singled out, tend to have inadequate holdings in scientific fields.

In regard to initial shop inventories, top-quality research equipment need not be purchased outright, but can be rented at moderate cost. Fig. 2-3 shows equipment descriptions and rental fees for typical laboratory instruments. A complete listing may be obtained from:

Laboratory Instrument Mart
301 Erie Street
Chicago, Ill. 60611

Note, however, that equipment used for critical research and engineering work must be certified as to its operational accuracy. Certification and/or calibration may be obtained from standards laboratories in larger cities. Resistive, inductive, or capacitive standards used for calibration of your equipment are traceable to the National Bureau of Standards, Washington, D.C. Nonstandardized lab equipment is *not* acceptable to government agencies and their subsidiaries.

Digital Equipment Corporation Computer, Model PDP 8/I. With memory and memory extension control Model MC 8/IA and ASR 33 teleprinter and program package (rack mountable, slides included). Approximately 6 months old. Rent for 3 months for $840.00 per month, 6 months for $554.40 per month. Available for immediate shipment.

Digital Equipment Corporation Computer, Model PDP 8/I. With 4K memory and ASR teleprinter and program package (rack mountable, slides included). Approximately 6 months old. Unused. Rent for 3 months for $640.00 per month, 6 months for $422.40 per month. Available for immediate shipment.

SELECTED RENTALS (New Equipment)

Technicon SMA 12/60 Autoanalyzer.
18 Mos. $1,800/mo.   5 Yrs. $1,306/Mo.

Technicon Dual Channel Autoanalyzer.
1 Yr. $294/Mo.   5 Yrs. $157/Mo.

Coulter Model S Cell Counting System.
1 Yr. $1,140/Mo.   5 Yrs. $612/Mo.

Nuclear-Chicago Pho/Gamma III Scintillation Camera.
1 Yr. $1,345/Mo.   5 Yrs. $871/Mo.

Picker 5″ Magniscanner.
1 Yr. $689/Mo.   5 Yrs. $446/Mo.

Digital Equipment PDP-8/LA Computer.
1 Yr. $280/Mo.   5 Yrs. $173/Mo.

Digital Equipment PDP-8/I Computer.
1 Yr. $422/Mo.   5 Yrs. $261/Mo.

Digital Equipment PDP-12C Computer.
1 Yr. $492/Mo.   5 Yrs. $304/Mo.

Hewlett-Packard 9100A Calculator.
1 Yr. $171/Mo.   5 Yrs. $91/Mo.

Olivetti Programma 101 Calculator.
1 Yr. $146/Mo.   5 Yrs. $78/Mo.

Beckman DU-2.
1 Yr. $139/Mo.   5 Yr. $75/Mo.

Beckman DB-G.
1 Yr. $108/Mo.   5 Yr. $58/Mo.

Beckman Ultracentrifuge Model L2-65B.
1 Yr. $367/Mo.   5 Yr. $197/Mo.

Perkin-Elmer Atomic Absorption Model 303.
1 Yr. $249/Mo.   5 Yr. $134/Mo.

Perkin-Elmer IR Spectrophotometer Model 137B.
1 Yr. $218/Mo.   5 Yr. $117/Mo.

Perkin-Elmer IR Spectrophotometer Model 237B.
1 Yr. $256/Mo.   5 Yr. $137/Mo.

Bausch & Lomb Binocular Microscope Model BB-352.
1 Yr. $40/Mo.   5 Yr. $25/Mo.

Courtesy Laboratory Instrument Mart

Fig. 2-3. Sample list of laboratory and research equipment for rent.

EXHIBITIONS: PATEXPO

Wholesaling of technical products is done in many ways. Some inventors became "wagon jobbers" who sold their wares out-of-the-car to retailers or, though less frequently, journeyed house-to-house. This is a time-consuming task, and large customer populations cannot be reached by this method.

Today, one of the best vehicles for introducing prospective buyers to inventor-made merchandise is the *patent exposition* (PATEXPO). The PATEXPO address is:

> PATEXPO
> Exposition Management
> International New Products Center
> 680 5th Avenue
> New York, N.Y. 10019

In a typical exposition of this type, the inventor's booth exhibit is seen by over 35,000 buyers and decision-makers in the patent-licensing field. A display booth 10 × 10 feet in size rents for $400; the largest rents for $600. Exhibitors must follow a code of exposition regulations, and are required to submit a $50 deposit at the time of application for booth space. However, these costs should be regarded as low, since inventors receive a tremendous exposure and good leads at the exposition. Truly spectacular inventions receive free press coverage and are noted by roving television reporters. PATEXPO is visited by representatives from all major corporations both here and abroad, NASA, the U.S. Department of Commerce, and foreign governmental organizations.

ARTISTS AND ADVERTISING

Shows like PATEXPO, mentioned above, should not be regarded as the only medium for successful display of inventions. Elsewhere, mostly in larger cities, inventors occasionally work with exhibiting artists, and the cost of renting gallery-type exhibition space is shared. These arrangements are highly beneficial, since artists like to blend their exhibit with technical innovations. Artists usually advance ideas on how to inject more liveliness and human appeal into the inventor's advertising material.

An example of this approach is illustrated in Fig. 2-4, which represents the basic artwork for drawing more direct attention to ELCO electrolytic capacitors. The actual picture of the electronic component, including text, may then be added as necessary. Also, if done in good taste, a pin-up model may be displayed for this purpose.

For best results, advertising and graphics should *always* be prepared by professionals. It is a frequent observation that, if descriptive material is prepared by the inventor himself, specific highlights attached to the invention tend to be buried in superfluous text, and people do not have the time nor inclination to wade through unneccessary prose.

Since it is difficult to describe, say, a complex machine in the limited space of an advertisement, prospective customers are invited to write for detailed information. The latter normally takes the form of a sales brochure, which may be combined with the equipment design philosophy and application notes. By contrast, a typical advertisement illustrates *dominant* highlights only.

MILITARY INVENTORS AND INVENTIONS

Both here and abroad, military invention has enjoyed a long and continuous history. In the United States, military-type technical innovations are encouraged by the three principal branches of Service—Air Force, Army, and Navy—and their subgroups.

Some of the earliest military inventions were extensions of civilian implements used in hunting. The crossbow, for example, came from the bow, and knives gave rise to swords. The emergence of sophisticated armies and warfare, in which inventive skills were pitched against one another, laid the foundations of modern weapons engineering. Modern amphibious assault vehicles, for example, found their beginnings in the crude river-crossing cars shown in Fig. 2-5.

Ultramodern armament, including nuclear devices, went through similar, but more accelerated, steps of evolution.

Weapons engineering and development normally are performed by contractors and/or engineering groups employed by governmental agencies. The average enlisted man or officer normally has no access to the research laboratories and massive resources required for prototype work. In fact, the only lab available to all on a military base normally is little more than a hobby-type photo laboratory. Or there may be a small shop where men can fabricate, also on a hobby basis, things such as model airplanes, a few sheet-metal items for automotive use, and the like.

Although such facilities might seem restrictive, they frequently do not hamper the development of ideas leading to the creation of advantageous methods or equipment systems. A good military invention might be, for example, an improved method for *washing* fighter aircraft. Or new ways may be found for rapid snow removal on strategic runways. In this latter case, the military inventor would probably consider equipment available for his purpose, such as a ramp trailer and spare Type J-47 jet engine. By mounting the engine on the trailer and directing its immense blast of heat against the snow, a highly efficient, useful, and practical invention could be made. However, to qualify this approach as a *genuine* invention under the terms of patent law, the military inventor probably would add deflectors and control surfaces to the basic concept in order to present a distinct improvement of the art.

Inventions made by military personnel must be formalized in accordance with military protocol. Fig. 2-6 shows a portion of AF Form 1280, used by the U.S. Air Force. This form is augmented by another form, AF Form 1279, entitled "Disclosure and Record of Invention." These two documents encompass the entire history and background

Fig. 2-5. Medieval amphibious armored car.

INVENTION RIGHTS QUESTIONNAIRE

INSTRUCTIONS (Submit four (4) copies of this form)

1. Under Executive Order 10096, 23 January 1950, as amended by Executive Order 10930, and AFR 110-8, whenever an invention is made by an officer, enlisted man or civilian employee of the Department of the Air Force, it is necessary to determine the rights in the invention as between the Government and the inventor. There are three ways in which rights may be determined: (1) the inventor may be entitled to all rights and the Government to none (and hence the inventor need sign no document giving any rights to the Government); (2) the Government may be entitled to a license permitting it to use and practice the invention and the inventor entitled to all other rights (and hence the inventor signs a license to the Government); (3) the Government may be entitled to all rights and the inventor to none (and hence the inventor signs an assignment to the Government).

2. Separate and distinct from the determination of rights, and even though it may appear that the inventor is entitled to all rights in the invention, the inventor may sign a license permitting the Government to use and practice the invention in return for which the Government will prosecute an application for a patent on the invention at no expense to the inventor, provided the Government is sufficiently interested in the invention.

3. If the inventor desires voluntarily to assign all rights in the invention to the Government, he should check 1(h) below as applicable, complete the answers to question 1 of this questionnaire and sign in the place provided at the end of question 1. The remaining questions then need not be answered.

4. If the inventor does not desire to voluntarily assign all rights in the invention to the Government, it is necessary that all questions be answered completely. The determination of rights in the invention depends upon the facts and circumstances under which the invention was made. In almost every case, a failure to provide sufficient information works to the disadvantage of the inventor. Every question must be answered. If more space is needed to fully answer any question(s) use blank sheets, identify the question(s) and attach. Print or type all answers.

5. Co-inventors should use separate forms and attach. Similar information may be referenced.

SECTION A (To be completed by the inventor)
BASIC DATA

1.

a. BRIEF TITLE OF INVENTION

b. NAME OF INVENTOR

c. JOB TITLE AT TIME INVENTION WAS MADE

d. GRADE AT TIME INVENTION WAS MADE

e. COMPLETE NAME OF ORGANIZATION (Including, as applicable, unit, section, branch, division, department, laboratory, base, center, area, command)

f. THE INVENTION WAS MADE (Check one)

☐ PRIOR TO ☐ SUBSEQUENT TO

23 JANUARY 1950

g. BRIEF SUMMARY OF INVENTION

h. KEEPING IN MIND THE STATEMENTS OF PARAGRAPHS 1 AND 2 OF THE ABOVE INSTRUCTIONS, CHECK ONE OF THE FOLLOWING:

☐ I DESIRE TO ASSIGN TO THE UNITED STATES GOVERNMENT THE ENTIRE RIGHT, TITLE AND INTEREST IN AND TO THE ABOVE IDENTIFIED INVENTION. (FOR THE RECORD, I HAVE BEEN ADVISED OF THE FULL IMPORT OF AN ASSIGNMENT AND AM AWARE OF MY RIGHTS OF APPEAL UNDER EXECUTIVE ORDER 10096, AS AMENDED BY EXECUTIVE ORDER 10930.)

☐ I DESIRE TO LICENSE THE UNITED STATES GOVERNMENT IN THE ABOVE IDENTIFIED INVENTION UNDER 35 U.S.C. 266 AND/OR TO EXECUTE A LICENSE IN SAID INVENTION TO THE GOVERNMENT IN ACCORDANCE WITH 1(b) OF EXECUTIVE ORDER 10096 AS AMENDED BY EXECUTIVE ORDER 10930.

☐ I DO NOT DESIRE TO GIVE THE U. S. GOVERNMENT ANY RIGHTS IN AND TO THE INVENTION.

DATE

SIGNATURE OF INVENTOR

MAKING THE INVENTION

NOTE: The making of an invention generally involves its conception or discovery followed by a series of acts which establish the correctness or operativeness of the idea. Depending upon the nature of the invention, these acts may involve any one or all of the following; the making of sketches, drawings, written descriptions, the making and testing of a model, the carrying out of a process, or the production of a composition of matter.

2. BEFORE THE INVENTION WAS PHYSICALLY TRIED OUT OR PRODUCED IN MODEL OR WORKING FORM OR A COMPOSITION OF MATTER PRODUCED, WERE THE ESSENTIAL ELEMENTS OF THE INVENTION IN ITS OPERABLE AND PRACTICABLE FORM FULLY DISCLOSED IN WRITTEN DESCRIPTION, SKETCHES OR DRAWINGS IN SUCH A MANNER THAT THE INVENTION COULD BE PRODUCED OR PRACTICED FROM THEM WITHOUT THE EXERCISE OF ANY FURTHER INVENTIVE SKILL BY A PERSON WHO IS SKILLED IN THE FIELD TO WHICH THE INVENTION RELATES? ☐ YES ☐ NO. IF THE ANSWER IS "YES", GIVE THE DATE SUCH DESCRIPTIONS, SKETCHES OR DRAWINGS WERE COMPLETED?

3. WAS A FULL SCALE WORKING MODEL OR A PROTOTYPE OF THE INVENTION MADE; OR, IF THE INVENTION IS A PROCESS, WAS THE PROCESS TRIED OUT; OR, IF THE INVENTION IS A COMPOSITION OF MATTER, WAS A COMPOSITION PRODUCED?

☐ YES (State date _____) ☐ NO

4. IF A MODEL OR PROTOTYPE WAS MADE AND TESTED, A COMPOSITION PRODUCED OR A PROCESS CARRIED OUT, WAS IT DONE BECAUSE: (Check)

	YES	NO
a. IT WAS DESIRED TO TEST THE OPERABILITY OR PRACTICABILITY OF THE INVENTION		
b. IT WAS DESIRED TO TEST THE USEFULNESS OF THE INVENTION TO THE GOVERNMENT		
c. OTHER REASON (Explain)		

AF FORM 1280 APR 63   PREVIOUS EDITION OF THIS FORM WILL BE USED.

Fig. 2-6. AF Form 1280 (page 1 of 4 pages).

of a military-type invention—by whom and where it was made, government time and supplies used, and so on. If necessary, an appropriate security classification is attached to the inventor's submission. Notarization is provided by the provost marshall's office.

A soldier, airman, marine, or sailor is an employee of the government—a division thereof, a given branch of Service, his immediate employer. Thus, if the military establishment receives an invention of exceptional merit and promise, government funds set aside for the purpose of rewarding innovative personnel are involved. In regard to the U.S. Air Force, inventors receive $50 for each favorably evaluated invention ($50 to each inventor in the case of joint invention), and the inventor or inventors receive $100 if and when a patent is issued. If the invention is of sufficient importance to the Air Force, its originators receive a much larger amount. Several awards of $5000 have been made. It goes without saying, perhaps, that the military likes to promote its people accordingly. The consequent increase in pay is an additional benefit to inventiveness.

General and Selected Problems

1. Peak-Holding Indicator

Instrumentation with peak-holding indicators is desired in many industrial areas. Unfortunately, owing to the expense and complexity of devices available, meters sometimes are not purchased, and risks are taken instead.

An approximate solution is projected in Fig. 3-1. The peak-holding system avails itself of a self-generating, logarithmic photoelectric cell. With excitation provided by a fixed light source, a shutter is activated, through a retainer ring and pin, by the pointer of the master meter. Since the cell current increases in proportion to the magnitude of the illumination reaching the cell, power is available to overcome the spring tension of the D'Arsonval movement (M). Diode D1 prevents rapid back-flow; capacitor C1 affords both damping and energy storage.

The pointer of the master meter is free to move in directions A and B. The peak-holding system, owing to unidirectional pull by the pin, is restricted to direction A only, i.e., toward peak maxima.

Improvements attendant to the problem reside in areas of zero-drift properties, fast attack time, and attenuation of system leakage for stable, long-term performance.

2. Engine Silencers

Advanced silencers are needed for appliances powered by gasoline engines. Loudness should be attenuated to +15 dB, but without impairing engine performance.

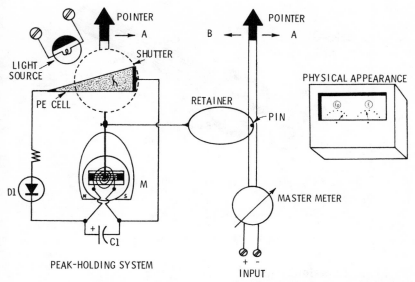

Fig. 3-1. Peak-reading voltmeter with hold feature.

3. Onion Peeler

A simple onion peeler, permitting homemakers to cut and dice the vegetable without smarting eyes, is needed. Many methods used so far remove too many of the outer layers of the onion or, at the opposite extreme, peel incompletely. This is an excellent market!

4. Slotted Socket Wrench

There is a long and continuous need for a rugged, inexpensive ratchet wrench for tightening nuts on long bolts. Since common sockets have insufficient depth, a special slip-on arrangement is required.

As shown in Fig. 3-2, the rotating nut-grip has been slotted to permit the bolt proper to pass through. The race must be fully reversible and permit high-torque operations. A nut-grip change feature is desired to accommodate different nut geometries and sizes.

5. Glass Drill

A rapid method of drilling glass for electronic applications without danger of breakage is needed. Preferred sizes are those of potentiometer shafts, toggle switches, screws, and connectors.

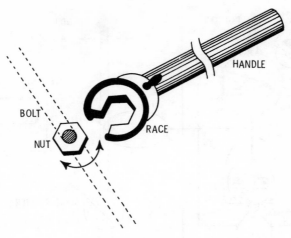

Fig. 3-2. Slotted slip-on wrench.

6. Meat Reaction Paper

There is need for an edible, chemistry-based reaction paper to be packed with meat products in display wrappings. By chemical reaction with aging ground beef, for example, the paper would indicate freshness and threshold of spoilage.

7. Plastic Umbrella

A plastic umbrella that can be guaranteed not to turn inside out during storms is needed. Special restrainers and locking shafts should bring results.

8. Clamp for Elevator Car

Better clamping methods are needed for attaching cables to elevator cars. Contemporary techniques lead to crystallization, causing cables to break. Flexible joints are required.

9. Oil-Well Drill Recovery Tool

There should be a market for a means of recovering drills lost in oil wells. Current "fishing" methods are inefficient and frequently make it necessary to abandon holes obstructed by lost tools.

10. Burnishing Check Writer

A need exists for a simple electric method of preventing tampering with checks. Check printers are too costly for many establishments.

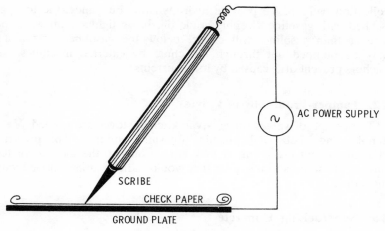

Fig. 3-3. Electric burnishing pencil for checks.

Fig. 3-3 gives one approach. Conductive check paper is placed on a flat ground plate, and entries are burnished into it by means of an ac power supply and scribe.

11. Oil-Spill Collector

Effective techniques are needed to prevent spillage from oil wells drilled offshore. Floating crude oil is lethal to marine creatures and triggers immense economic losses if oil slicks enter beach areas set aside for recreation purposes. Fig. 3-4 suggests the use of a plastic

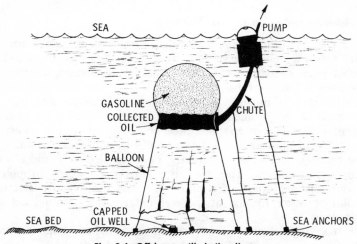

Fig. 3-4. Offshore spilled-oil collector.

collection balloon. To provide hydrodynamic buoyancy, the top part is filled with gasoline. Seeping crude oil, being lighter than water but heavier than gasoline, will collect below the gasoline horizon and may be pumped out through the chute by external machines. Sea anchors prevent drift caused by local currents.

12. Tamper-Proof Stove Valves

Tamper-proof valves for gas-type kitchen stoves are needed. Valve handles and knobs invite turning by small children. The preferred invention might use a master lock known only to the mother, or feature an electrical alarm system that would indicate unauthorized valve movements.

13. Noncracking Concrete

Costs of road maintenance could be reduced sharply by providing state and municipal customers with noncracking concrete.

14. Street Lighting

A method for changing the color of street lighting without multiple lamps and/or filters is needed. Lighting attracts insects, which spread to nearby homes. Yellow is but one color that provides sufficient illumination after rush hours; many insects are color-blind to yellow. White light may be used during periods when the traffic density is high.

15. Electroendosmosis for Plowing

The tractor is the farmer's modern horse. Unfortunately, for economic reasons, vehicular speed is subordinated to the magnitude of drawbar load. Usually, when higher speeds are desired, a larger and more powerful tractor must be purchased.

However, it is possible to reduce the draft in plowing by electrical methods. Experimentally tried in the early 1920's, the phenomenon of *electroendosmosis* was used in conjunction with moist soil (Fig. 3-5). Because of the negative charge of soil colloids, water will move through moist soil toward the negative electrode under action of an electric current. Therefore, if a current is passed through the soil and the moldboard of a plow serves as the negative electrode, the film of water formed at the soil-metal surface should act as a lubricant.

Draft savings of at least 30 percent are sought. About 220 watts of dc power are required for a soil resistance of 100 ohms.

Courtesy A. Williams

Fig. 3-5. Use of electroendosmosis for better draft.

16. Polisher/Buffer

A special machine for polishing and buffing of odd-shaped metal articles and polymer castings (plastics) is needed.

17. Readout for Electronic Calculator

Inexpensive, electronic desk calculators featuring a print-out mechanism are needed. The problem is considerably more complex than it appears at first glance! Efforts have been made to use *Nixie*-type digital indicators to project images on special photosensitive paper; this approach tripled the cost of the calculator. Other attempts employed electrical burnishing methods whereby number patterns where formed by hot needles pressed against paper. New methods might use a rotating print-out head as is used in some IBM typewriters, with stop signals furnished by electromagnetics. This field is wide open for improvements.

18. Phono-Disc Compound

A new compound is needed for making stereo phonograph records that will retain their surface characteristics regardless of how often they are played by both new and inferior equipment (i.e., a pickup with a bad needle).

19. Can Safety Valve

Pressurized spray cans for household use are extremely dangerous, since explosions result from heating them. In cold climates, these cans frequently become nonfunctional. To overcome freezing of the contents, a special safety can is needed which, if equipped with a small safety valve, would permit heating without danger.

20. Film Overexposure Prevention

A photographic film that cannot be overexposed should be popular. Today, good photography involves the use of ASA or similar film ratings, and proper setting of camera f stops and shutter speeds on the basis of these ratings and the ambient lighting conditions. Films use different silver-based coatings for recording a latent image. Since the constituents change during the event of exposure, the overexposure prevention device might make use of this phenomenon.

21. Generator for Model Airplane Engine

Electric generators aboard model airplanes would be enjoyed by many hobbyists. Orthodox generators are bulky and require engine modifications. An intermediate solution is given in Fig. 3-6. The propeller includes a ferromagnetic sleeve. As the sleeved propellor turns past a permanent magnet adjacent to the engine housing, inductive pulses are generated in the winding around the magnet. The pulse train may be rectified, filtered, and made available to the servos, radio-control equipment, and the like in the plane. The generator

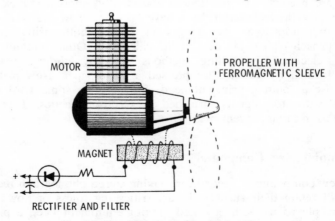

MOTOR

PROPELLER WITH
FERROMAGNETIC SLEEVE

MAGNET

RECTIFIER AND FILTER

Fig. 3-6. Electric generator for model airplanes.

would recharge one or more small batteries continuously, the latter providing high currents on demand.

22. Pipeline Monitor

Petroleum pipelines frequently carry different types of liquids, ranging from crude oil to high-octane gasolines. One train of liquid may be transported behind another. A rubber-type "ball" or similar insert may be used to separate these liquids. This insert must be removed at its destination.

However, because of their different densities, liquids produce different noises in their dynamic stream patterns. This phenomenon may be used as is suggested in Fig. 3-7. A small bypass tube is attached to the main pipe, and an acoustical transducer is fixed to the bypass path. If the liquid density changes, a different pitch will be produced. This pitch is frequency-evaluated by readouts which, in turn, might activate warning devices for the valve operator.

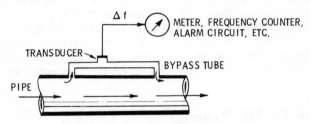

Fig. 3-7. Acoustical liquid discriminator.

23. Wire-Drawing Tool

There is need for a new method of drawing wires and cables. Current processes make use of diamond dies. Although lasers may be used for drilling diamonds rapidly, the expense of the basic diamond still remains high.

24. Multi-Image Storage Oscilloscope

A multi-image storage oscilloscope is needed for rapid data comparison. Currently, in order to preserve data when critical experiments are undertaken, graphic recorders must be employed. There is no reason, however, why some of these tasks could not be performed by the oscilloscope itself.

Storage-type scopes have been available for many years. Special phosphors and excitation grids are used for retaining an image on the

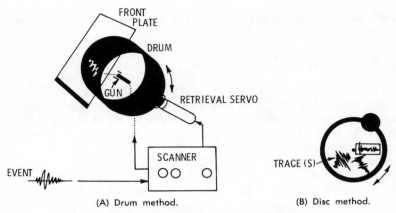

(A) Drum method.　　　　(B) Disc method.

Fig. 3-8. Multi-image storage oscilloscope.

screen after it has been written by the gun. If, as in Fig. 3-8A, the screen is manufactured in the form of a drum, a multitrace storage capability can be achieved. A disc method also may be used (Fig. 3-8B). The drum is driven by a retrieval servo and can display, at will, the various recordings for comparative purposes. Image-retaining charge structures must, of course, be arranged throughout the interior geometry of the drum. The servo motor operates in the vacuum of the CRT.

25. Rims for Mill Rollers

For reasons of economy, steel mills are seeking methods for reconditioning mill rollers (Fig. 3-9). The reconditioning process must eliminate material fatigue and re-establish the precision of the roller. Typical rollers produce steel sheets, welding rods, screw stock, and stock for seamless billets. Currently, due to lack of a suitable invention, used rollers are returned to the melt. The problem involves rimming techniques and induction heating to avoid crystallization of the metal.

26. Improved Lubricity of Water

Methods are needed to improve the lubricity of water. The improvement would reflect less friction in water-guiding pipelines, resulting in savings of pump power. Better lubricity also would improve the propulsion characteristics of ships, naval weapons such as torpedos, and the like. Concentrations of less than 100 parts per million of substances like polyethylene oxide and guar gum reduce friction between water and solid bodies by as much as 40 percent.

Fig. 3-9. Rollers for steel mills.

27. Inexpensive Boat Pilot

Model boats operated on ocean waters are difficult to see and to control by radio. This handicaps competition runs over great distances. A special autocompass is needed for pilot duties. However, because of high power requirements, conventional apparatus such as gyroscopes cannot be employed. Fig. 3-10 suggests a photoelectric compass arrangement. The compass at the bottom is a reference unit; the compass at the top is furnished with a small light bulb beneath its needle. A photoelectric cell is contained in the terminal connector

Fig. 3-10. Simple magnetic-type autopilot
for full-size or model boats.

of a fiber-type light guide. Thus, if course deviations occur, the cell receives changing light values that depend on the movement of the compass needle, in this case acting as a shutter. Cell currents may be used to activate the rudder, etc. A similar system may be used for small, manned boats operating on fresh-water lakes. This permits automatic guidance during trolling operations.

28. Washer for Fruit Vendors

Fruit vendors serving highway customers have need for a simple washing device. To our knowledge, fruits sold at open-air stands (typically cherries, apples, grapes, peaches, etc.) are washed in the presence of customers only in the state of Utah! Others wash their fruits elsewhere, usually away from stands. The frequent lack of business experienced by the latter group is based on psychological and sanitary rejections by prospective customers. Thus, an attractive and economical fruit washer for general use would benefit these vendors.

29. Golf Trainer

Beginners in golf would appreciate a club designed for heavy abuse and incorporating an alarm feature. The latter should sound if ground strikes are too heavy, i.e., a lot of dirt is moved rather than the ball.

30. Epoxy Pouch With Separators

Epoxy glues are superior to ordinary mending aids in many applications. Unfortunately, since they are sold in packages of two containers having different contents (glue and catalyst), rapid aging is unavoidable once the seals have been broken. A plastic epoxy bag (Fig. 3-11) would help to mitigate this problem. The bag contains

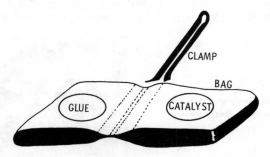

Fig. 3-11. Epoxy bag with separator pouches and clamp.

small, separated quantities of glue and catalyst. Once the separating clamp is removed, the substances can be mixed by kneading the bag.

31. Emergency Bridge

Flooded areas, with bridges washed out, demand immediate restoration of facilities for emergency and general vehicles. Immediate restoration of bridges has been a long-standing problem. Good rigidity is required for safety. The ancient assault bridge, shown in Fig. 3-12, offers some excellent features for this purpose. First, a gap is bridged by means of a beam system. Second, a prefabricated roadbed—here composed of folded planks—is cranked into the beam structure once firm ground has been reached. The final bridge may be diesel-powered and self-propelled.

Fig. 3-12. Principle of dated assault bridge.

32. Piano-Tuning Device

A simple instrument, preferably electronic, is needed to permit tuning of pianos without expensive professional aid.

33. Nonelectronic TV for Schools

The installation of TV studios in public schools is not widespread. Although self-contained student study carrels would bring enormous

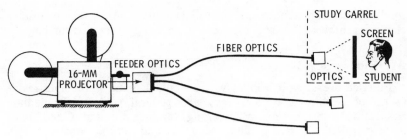

Fig. 3-13. Principle of nonelectronic television.

benefits to educational target objectives, the high cost of TV playback and production facilities prohibits acquisition.

The average educator does not appreciate the technological complexities attached to present-day TV. Therefore, in order to find a rewarding market in this open field, different methods of televising images must be investigated. These methods include nonelectronic, *optical* television.

Fig. 3-13 suggests one approach. A 16-mm film projector projects program material into a fiber-optic system. Optics at the terminal end then project the image onto student screens. Sound can be conveyed to study carrels by simple lines feeding loudspeakers. A simple mechanical shutter may be inserted between optics and screen to turn the image off. Different programs, originating from different projectors, may be displayed on the same student screen by shifting appropriate fiber light-guides in the path of the display optics.

An arrangement such as this makes fast learners no longer captive to slow and boring classroom situations. Technically, areas requiring inventive talents reside in transmittance properties associated with optics and efficiency of the carrel display system.

34. Water Premixing Device

A special premixing system for bath tubs is needed to permit the bather to regulate the water temperature before water enters the tub. This item might consist of a simple bypass tube which directs the water into the drain during the premixing process, and finally allows the water to flow through the regular faucet once a preset temperature has been reached. Such an invention is desired in hospitals and rest homes where invalids bathe themselves.

35. Fresnel-Type Mural Projector

In many cities throughout the United States, special wedding reception centers are operated by caterers. Proper interior decoration

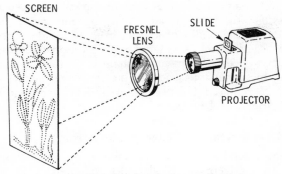

Fig. 3-14. Mural projection system.

of the center, including walls and ceilings, needs to be aligned with the personal tastes of the principals of the wedding party, not with whatever tastes the caterer might have. Here, special decorative schemes can be *projected* onto walls and ceilings of the center.

Fig. 3-14 suggests the technique to be used. From prepared projection slides, the image is directed through a Fresnel lens for purposes of proper optical geometry and focus. A ground-glass screen receives the image by rear projection. Projected slides might come from the couple's own collection or from the caterer's slide inventory.

36. Bathing-Suit Pocket

A special watertight pocket for bathing suits is needed for storage of keys or money. Zipper type methods are not satisfactory. The bulk of the pocket must be nil.

37. Emergency Trailer Brake Shoe

No efficient methods are available for stopping runaway house trailers on steep grades. If a passenger car functions as prime mover,

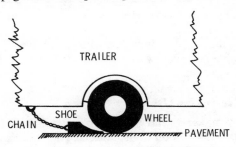

Fig. 3-15. Emergency trailer brake shoe.

lethal jack-knifing might result. A drastic approach is needed. In Fig. 3-15, a wooden shoe is suggested. During an emergency, the shoe is electrically or mechanically dislodged from its holder and directed beneath the trailer wheel. The latter will run atop the shoe, thereby losing contact with the pavement. A chain is paid out and retains the shoe. Owing to tremendous friction, the trailer will come to a halt. Two shoes are required for two-wheel trailers, and four shoes are required for four-wheel types.

38. Liquid-Conductance Tester With Toroids

Simple liquid-conductance testers have many applications. In the automotive area, drivers can be alerted to oil changes; in the domestic field, cooks can obtain concise information on the saltiness of soups and semiliquid foodstuffs. Aside from involved ohmmeter-type methods and their disadvantage of electrolysis, a scheme such as the one shown in Fig. 3-16 can be employed successfully. Here, an ac bridge incorporates two toroidal coils, fully encapsulated in plastic, as sensing elements. The reactance of these coils is determined by the flux components directed through the center design axis. Thus, if the ambient flux field is enhanced or attenuated by conductive constituents in the liquid (carbon residue in the case of motor oil), the bridge will unbalance and deflect the meter accordingly.

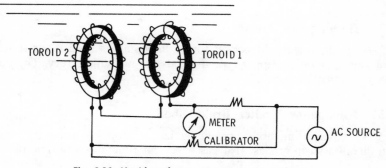

Fig. 3-16. Liquid-conductance tester uses toroids.

Aside from the applications mentioned above, the instrument would be valuable to those who have to restrict their salt intake for medical or dietary reasons.

39. Ribbon Re-Inking Device

A real money-maker would be a simple attachment for typewriters that would re-ink the ribbon as it is used. The inker must not smear the paper, and it must give uniform ink saturation.

40. Camera Gyroscope

An efficient accessory is needed to steady a camera while pictures are being taken. Expensive gyroscopic units use electricity and have too much weight. A spring-driven device would be appreciated by both movie and still photographers.

41. Follow-Object TV System

When video tape recordings of children are made with school-owned equipment, the subjects give exceptionally natural performances if recording takes place without the presence of a camera operator. To that end, especially where nonrepeatable experiments are involved, fully automated camera systems having follow-object features are desired. Special pan/tilt-head combinations operated by an "outside" technician could be used, but such solutions are much too involved and expensive.

The desired follow-object feature, which also could be applied to industrial and space situations, might be based on a simple lock-on device carried by the subject. Making use of received guidance information, the camera would maintain continuous homing until directed otherwise. This is a complex problem, but it has a good market potential.

42. Self-Extinguishing Cigarette

For prevention of forest fires, there is need for a cigarette designed with self-extinguishing features. Since the glow of a cigarette is based on oxidation processes, the problem might be conquered on this basis. Both the packing densities of tobacco and the paper used need to be evaluated.

43. Train-Journal Monitor/Recorder

Hot journals on both freight and passenger trains can lead to fire. Less drastic effects are friction and consequent waste of locomotive power. Journal failures may be caused by material fatigue, lack of lubrication, or obstructions. Many failures occur on open plains where no infrared-type detector stations are installed at given intervals, so heat goes unnoticed. To permit corrective action by train personnel, a portable heat monitor could be deposited on the ground by the locomotive engineer on slow grades and with the train in motion. Thus, as cars pass the monitor, heat anomalies may be radioed to the caboose and/or locomotive. The monitor could be picked up again by a crewman on the last car (the caboose in the case of freight trains).

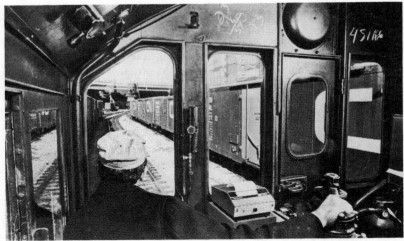

Fig. 3-17. Telescripter in diesel locomotive.

For the monitor to be effective, a car identification or counting feature must be available. Further, it is desirable that anomalous data be presented in a *printed* manner for logging purposes. Here, electronic compatibility with train telescripters is mandatory. The journal monitor might therefore transmit signals which are acceptable to the existing telescripter translator. Five-bit Baudot, 8-bit ASCII, or Hollerith codes are acceptable.

Fig. 3-17 illustrates the engineer's position; the telescripter is placed next to the locomotive throttle. No superfluous electronics can be tolerated because the engineer's main attention belongs on the road ahead. This is a ready and lucrative market.

44. Tank Purging System

A manual or semiautomatic device is needed for purging automobile tanks of water condensates and dirt.

45. Grease Collector for Ovens

Housewives would appreciate a grease run-off pan for ovens. As shown in Fig. 3-18, the item could feature a tilted meat holder and spout. The grease runs into the spout and is collected externally. A special oven, or oven attachment, is required to accommodate the exit spout.

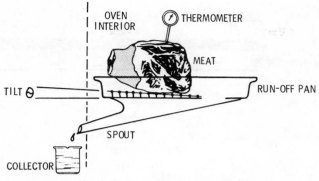

Fig. 3-18. Grease run-off for ovens.

46. Insect Suppressant for Food Plates

In tropical regions, it frequently is impossible to leave food standing on a table without inviting insects. An insect suppressant is needed. It might take the form of an adhesive tape impregnated with an odorless insect repellant. Tape can be furnished in a tear-off type plastic container and be applied to the bottom of a dish, plate, pot, or other dining utensil. This product also is needed by campers.

47. Vibratory Snow Remover

A self-powered appliance to remove compacted ice and snow layers from sidewalks would be appreciated by home and store owners. This appliance might have an oscillating or vibrating knife energized by electricity. An ice-melting feature could be incorporated.

48. Pipe-Noise Damper

Noisy water pipes can be a nuisance. To attenuate the transmission of noise generated by flowing water to walls and other structures, silence could be ensured with the arrangement shown in Fig. 3-19. Here, the water carrier is mounted coaxially in the center of the ducting channel.

49. Nonrecoil Satellite Weapon Systems

An attack-type space satellite is needed for defense purposes. The unit might feature a laser gun to disable enemy satellite electronics or, as shown in Fig. 3-20, it might use solid projectiles. Here, pro-

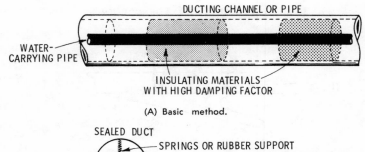

DUCTING CHANNEL OR PIPE

WATER-CARRYING PIPE

INSULATING MATERIALS WITH HIGH DAMPING FACTOR

(A) Basic method.

SEALED DUCT

WATER PIPE

SPRINGS OR RUBBER SUPPORT

VACUUM (IN LIEU OF DAMPING MATERIALS)

(B) Alternate method.

Fig. 3-19. Noise-free water pipe.

jectiles with explosive tips may be fired by an electromagnetic gun or accelerator (Fig. 3-20A). Two guns are required to equalize recoil effects. While the explosive projectile is being fired, a dummy projectile of equal weight must be ejected at the same velocity at the opposite end of the gun structure (Fig. 3-20B). Target positioning of the attack satellite may be afforded by means of a captive liquid conductor, such as mercury. If exposed to ac fields, the mercury moves in a motor tube and imparts a rotational component to the equipment system (Fig. 3-20C).

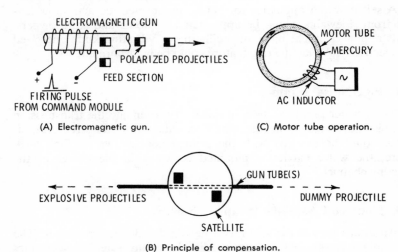

ELECTROMAGNETIC GUN

POLARIZED PROJECTILES

FEED SECTION

+

FIRING PULSE FROM COMMAND MODULE

(A) Electromagnetic gun.

MOTOR TUBE

MERCURY

AC INDUCTOR

(C) Motor tube operation.

GUN TUBE(S)

EXPLOSIVE PROJECTILES

DUMMY PROJECTILE

SATELLITE

(B) Principle of compensation.

Fig. 3-20. Attack satellite uses recoil-compensated electromagnetic gun.

50. Speech Trainer for Deaf Children

Speech training for deaf-mute children is complicated by the fact that the handicapped child cannot hear himself talk. Apparent muteness might be a result of this condition. However, if even a very small auditory sense exists, electronics may be used to furnish the child with speech "models" to repeat. Typical equipment for this application might use continuous tape loops, but would be heavy and bulky. Equipment of this type tends to be very expensive, which precludes home use. Thus, in order to afford children a training aid while away from the speech clinic, a simple instrument such as that shown in Fig. 3-21 might be evolved.

The heart of the apparatus is a magnetic recording disc driven by a dc motor. Within angular displacements of 90 degrees each, the disc is equipped with recording, listening, and erasing heads. A special track, not subject to erasure, might contain the speech model. Then, as the disc rotates, the child tries to match his articulation with that on the model track. Since playback is immediate, a learning situation has been established. The child's sound track will be erased upon passing the erasing head.

This field is wide open for improvement. Additional information is contained in the following two technical articles:

L. G. Lawrence, "Electronics for Speech and Hearing Therapy," *Electronics World,* September 1967, p. 44.

L. G. Lawrence, "Automatic Diplexers for Voice Communications," *Radio-Electronics,* September 1968, p. 48.

51. Repair of Ship Propellers

New techniques and methods are needed for repair or replacement of ship propellers on the high seas. If, as shown in Fig. 3-22, fractur-

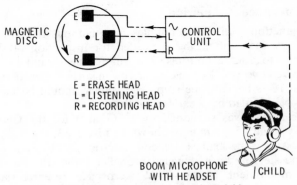

E = ERASE HEAD
L = LISTENING HEAD
R = RECORDING HEAD

Fig. 3-21. Articulation repeater for deaf children.

Fig. 3-22. Repair of ship propeller.

ing occurs, sleeve material may be placed over the fracture and riveted by divers. Welding is another expediency. None of these approaches is satisfactory, since high-torque operations tear the sleeving or rip welding seams.

52. Maintenance Submarine

In conjunction with the techniques and methods discussed above, a simple, two-man submarine is needed for inspection of ship hulls and damage to double bottoms. The submarine should be powerful enough to remove small obstacles or, by means of directional explosives, clear reefs beneath the ship. The vehicle should be inexpensive enough to be afforded by ocean freight lines.

A basic design was suggested by E. Goubet for the German imperial navy some 75 years ago. Shown in Fig. 3-23, the boat featured an air-pressure type ballast system, electric propulsion, a torpedo holder, and steering by a swiveled propeller rather than a rudder. A modern embodiment is the LTV Aerospace research submarine shown in Fig. 3-24.

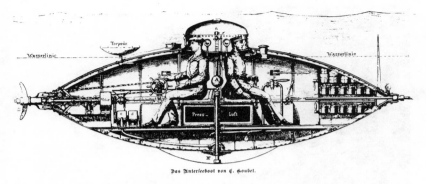

Das Unterseeboot von E. Goubet.

Fig. 3-23. "Unterseeboot" by E. Goubet.

Basic problems associated with inspection-type submarines are power and endurance with a minimum of weight. For 30 shaft horsepower (shp), up to 30 hours endurance, silver-zinc batteries may be employed. However, if additional power is required, together with breathing atmospheres, electricity may be fed by marine cables to the submarine. The latter might carry quick-disconnect systems and powered cable winches to take up slack.

53. Thermocompression Bonding Techniques

In the manufacture of integrated circuits (IC's), superior bonding techniques are required. Advanced inventions are needed for providing ultratight bonding of chips to mounting plates, yet without losing any of the required insulation properties. Such methods would afford

Fig. 3-24. Research submarine.

Courtesy LTV Aerospace Corp.

Fig 3-25. Flame-off torch for cutting gold wire during IC manufacture.

Courtesy Tempress Research Co.

much higher power-handling capacities of IC's and give a distinct competitive advantage.

Some of the processes attached to the manufacture of composite IC's are demonstrated in Fig. 3-25. Here, a flame-off torch is used for cutting minute (as small as 0.0007 inch outside diameter) gold wire. A ball for nail-head bonding is formed in a smooth flow of hydrogen through the torch orifice. The torch is tipped with a highly polished sapphire insert to ensure that size and geometry of the flame remain constant.

Pioneered by the Tempress Company, thermocompression lead-bonding capillary tools are now manufactured from tungsten carbide instead of glass. At the top of the picture (Fig. 3-25), ultrathin wire is inserted into the tool and emerges at its lower tip.

A typical TO-5 IC package involves many different manufacturing operations. However, metallic headers are superior to plastics. Invention in this area might revolve around special etching techniques and positive suppression of electrolysis. Thermoelectric cooling of the IC mounting plane would also be desired.

54. Inspection of High-Tension Lines

New, more positive methods are needed for both electrical and mechanical inspection of high-tension lines. High-voltage creepover can be detected by parabolic microphones, since the event produces noise. Visual inspection of lines is afforded by helicopter. However, only a casual inspection is possible. The danger of becoming entangled in the line(s) is always imminent.

Methods of more direct, vehicular line inspection were suggested at the turn of the century. This concept involved a bicycle-type ve-

Fig. 3-26. Line inspection by manned, self-propelled bicycle.

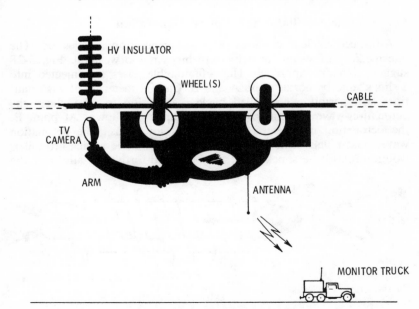

HV INSULATOR

WHEEL(S)

CABLE

TV CAMERA

ARM

ANTENNA

MONITOR TRUCK

Fig. 3-27. Self-propelled line inspection vehicle.

hicle (Fig. 3-26), self-propelled by electricity taken from the line. Although this approach might appear obscure, it could be used in the line-inspection apparatus suggested in Fig. 3-27. Here, a wheeled inspection device is hung onto the line (possibly by a crane). The battery-powered vehicle contains a TV camera mounted on a flexible arm, and, by fm-TV, signals are transmitted to the monitor truck below.

Continuous operation could be obtained by inventing a device that permits the vehicle to go past insulators. Insulators, as can be seen in Fig. 3-27, form a barrier to continuous travel. In Fig. 3-26, since the wheels ride above the insulators, motion without stop is possible. The new invention should aim at vehicular passing of line insulators since, owing to the costs involved, it is unlikely that special carrier cables can be added to existing installations.

55. Grease-Dissolving Kitchen Drain

A kitchen drain that features a grease-dissolving attachment is needed. Since grease melts by application of sufficient heat, an electrical heating pad may be employed. The disadvantage of using chemicals for the same purpose is that they are polluters and are corrosive to metal-based plumbing.

56. Nuclear Oscillator for Power Generation

Advanced nuclear methods of power generation are desired. The intermediary of steam for driving turbines is too wasteful. Fig. 3-28 suggests but one approach. Here, fissionable gases are injected into a shock tube by nozzle pairs A and B. Careful metering of gas quantities will permit a controlled nuclear explosion to take place; the detonation-wave plasma travels in the direction shown. At point B, the detonation is repeated by fresh gas injections. The detonation wave thus travels back to point A, etc. Now, since a pick-off coil is wound around the shock tube, the back-and-forth oscillations of the

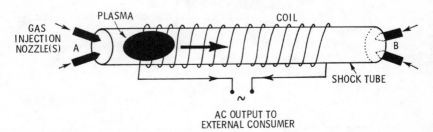

Fig. 3-28. Shock-type nuclear power oscillator.

Fig. 3-29. Eight-point explosion simulates oscillator detonation wave.

plasma induce a current by induction. The principle is quite similar to that of a permanent magnet entering and exiting a common solenoid. In the case at hand, alternating current will be obtained.

Note that explosions of any kind, including chemical ones, produce uncontrolled shock waves. Once ignition of a susceptible medium has occurred, a high-speed detonation wave results, as shown in Fig. 3-29 (recorded at 600,000 pictures per second on Anscochrome D/200 film exposed at 0.6 μs per frame).

In view of the above, the essential features of the invention will pertain to some control of powerful shock waves, cooling of the shock tube, dc excitation of the plasma, and control apparatus for metered gas injection.

57. Scientific Detection of UFO's

In his introduction to the controversial book *Scientific Study of Unidentified Flying Objects* (the so-called "Condon Report"), *New York Times* columnist Walter Sullivan wrote these words: "If, as many people suspect, our planet is being visited clandestinely by spacecraft, manned and controlled by intelligent creatures from another world, it is the most momentous development in human history. . . ."

The UFO scene has been with us since the day in 1947 when Kenneth Arnold was flying a private plane near Mt. Rainier, Washington, and reported seeing a group of objects of unknown origin flying in line. A newspaper reporter coined the term "flying saucer."

Does this novel field offer opportunities for invention? Yes, if in a somewhat restricted sense. Invention can cater to the needs of both private and academic observers by providing sophisticated instruments to suit their needs. To that end, and to be fair, it is necessary to alert customers to the complete history of "ufology."

The Condon Report may be regarded as a basic reference. The tenor of the book is pro and con, though mostly negative on this phenomenon. Next, it is interesting to examine ancient artifacts and legends which seem to imply extraterrestrial visits in ages gone by. The following information is reported without biased comments:

Fig. 3-30, an aerial photograph covering a region of more than 1000 square miles, shows mysterious markings on the stone-covered pampa near Nazca, Peru. The lines, geometrically derived, probably date back to ancient times and range over mountains, valleys, and passes. A similarity to modern-day landing fields may be noted. Legends relate that this massive amount of work was intended to signal aerial "bird deities."

Fig. 3-30. Aerial photo of stone-covered pampa near Nazca, Peru.

Fig. 3-31. Stone relief in Pyramid of Inscriptions, Palenque (southern Mexico).

Fig. 3-31 shows a stone relief from the ancient city of Palenque in southern Mexico. The relief is cut into the tomb plate located in the Pyramid of Inscriptions and was discovered by the Mexican archeologist Alberto Rusz in 1935. The relief is dated 900 A.D. The design is implied to be an image of the Toltec god Quetzalcoatl and his transportation. When the relief is examined closely, it is possible to conclude that the god is riding some sort of vehicle, with flames or gases exiting to the extreme right. Also, his hands are engaged in steering operations.

The above artifacts, available for direct examination today, seem to support numerous events of extraterrestrial visitations. However, just how much validity can be attached to this "evidence" remains to be seen.

Current UFO-directed investigation methods involve concepts of photogrammetry, as well as geomagnetic, radiological, and optical methods. Fig. 3-32 shows a once-famous UFO picture that has been analyzed by photogrammetric techniques. By subjecting the photo to measurements, specifically by analyzing shadows and distances, it was found that the object was in the foreground and only two feet in diameter—not 20 feet as claimed!

Fig. 3-33 illustrates some instruments in the family of UFO detection gear. Left to right, they are: search-coil magnetometer, radiometer, and Newtonian reflector telescope. The Newtonian instrument is based on optics and permits visual detection and photography. The radiometer, here equipped with an aluminum parabolic reflector, is intended to collect UFO radiation properties, if any. The magnetom-

Fig. 3-32. Alleged UFO picture.

eter is a true geophysical instrument and is highly sensitive to both pulsating and ambient magnetic fields.

The operating principle of the magnetometer is delineated in Fig. 3-34. As shown in the diagram, a high-impedance coil is motor-driven

Fig. 3-33. Principal UFO detectors used for field investigations.

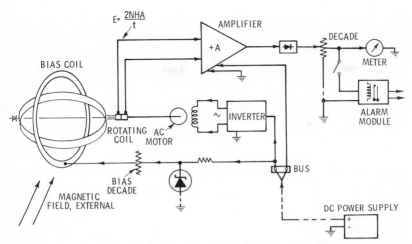

$$E = \frac{2NHA}{t}$$

AMPLIFIER

BIAS COIL

+A

DECADE

METER

ROTATING AC
COIL MOTOR

INVERTER

ALARM
MODULE

BIAS
DECADE

BUS

MAGNETIC
FIELD, EXTERNAL

DC POWER SUPPLY

Fig. 3-34. System electronics of search-coil magnetometer.

above the surface of the earth. Thus, as the coil cuts the geomagnetic flux, a voltage is generated. This voltage is picked up by slip-rings and fed to an amplifier. After amplification, the signal is applied to a decade system and read out by a meter or, in the case of auditory monitoring, by an alarm module. The metered display is in terms of magnetic *gammas* (1 gamma, γ, equals 10^{-5} oersted). Owing to the excellent magnetometer resolution of better than 1.5 gammas, magnetic anomalies existing or occurring adjacent to the magnetometer station will be detected without fail.

Fig. 3-35 shows another custom instrument invented for impartial UFO investigators. Created in Switzerland and named Wirbelstromgerät (eddy field sensor), the apparatus was intended to detect rotating induction fields generated by "flying saucers."

Fig. 3-35. Eddy-current UFO field sensor.

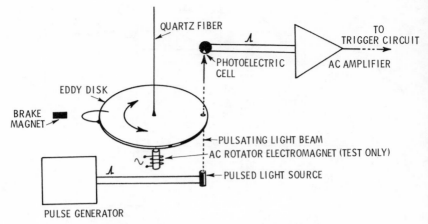

Fig. 3-36. Diagram of eddy-current UFO field sensor.

Fig. 3-36 shows the principle of the unit in Fig. 3-35. An aluminum eddy disc is suspended from a quartz fiber and held in a quiescent mode by the slight braking action of a magnet assembly. A pulsating light shines through a hole in the disc and is received by a photoelectric cell. Thus, should the cell-connected amplifier receive no more signals due to the rotation of the disc by an external force, relays will drop out and activate alarm circuits. An ac-type pulsation was used in order to obtain convenient signals for transmission over telephone lines.

The above represents only a sample of a unique art. But this field is important in its own way. Purpose-tailored inventions can divorce ufology from its occult background and, by means of verifying instruments, attach it to traditional physics. The markets are isolated, but rewarding once found.

58. Three-Dimensional Orrery

Simple, accurate techniques are needed for presenting astronautical data in three-dimensional form. The ideal device would correlate planetary versus spacecraft positions, give immediate information on distances traversed, and the like. Current methods deal with local problems only (moon/earth-space) and are not suitable for true interplanetary applications.

Refined versions of the 18th century *orrery* (Fig. 3-37) might advance solutions. Being the forerunner of the modern planetarium, the apparatus illustrated relative positions and motions of bodies in the solar system. A computerized input system is required.

Fig. 3-37. Three-dimensional orrery.

59. Tidal Power Plants

In order to counteract environmental pollution, new techniques for power generation are desired. Nonthermal methods are preferred. This problem invites a consideration of natural phenomena, typically as associated with water and its motion during cycles of the tide.

Fig. 3-38 shows a flour mill installed during medieval times by the city of Dover, England. Using hinged pressure planes and a reser-

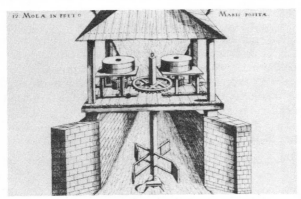

Fig. 3-38. Tide mill.

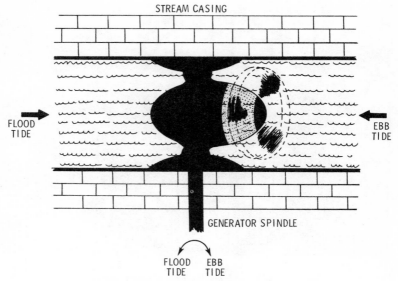

Fig. 3-39. Tidal hydroelectric system using Kaplan turbine.

voir, the mill was able to operate continuously. Rotational reversals of the runner made it necessary to engage millstones in the main drive alternately.

Preliminary experiments have indicated that high-pressure operation is afforded during ebb tide, water being furnished from a reservoir. One or more generators can be driven as in Fig. 3-39 with a considerable efficiency. The illustration implies flood-tide operation as well, but at reduced efficiency. The turbine system might incorporate a generator within its housing or, better yet, drive an external unit by means of a gearhead and a spindle (or shaft).

The ideal reservoir would resemble that of a traditional hydroelectric station such as Hoover Dam (Fig. 3-40). Here, intake towers are provided to direct water to turbine groups. Such an arrangement affords excellent redundancy in case of turbine failures.

In the case at hand, since corrosive salt water is involved, special plastics or other coating materials are needed to prevent damage caused by electrolysis.

60. Cable Security Loop

A reliable method to secure privacy in cable-carried communications would be a gainful invention. Tampering in this area frequently is accomplished by cutting through cable wrappings and tapping the wires within the cable proper.

Fig. 3-40. Water storage behind Hoover Dam.

Fig. 3-41 illustrates an approximate approach. Here, a bifilar guard cable has been wrapped around the main cable. Thus, in order to have access to the interior of the main cable, it is necessary to cut through the guard cable first. This act will trigger the control system connected in series with the bifilar line.

The reason for the bifilar arrangement is as follows: Many communications cables feed both low- and high-impedance circuits and are generally susceptible to induction currents. A bifilar guard cable cancels its own magnetic field by virtue of its structure and does not contribute interference.

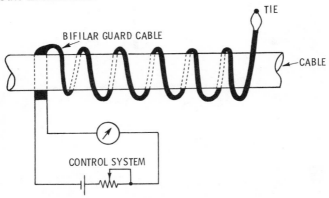

Fig. 3-41. Security cable with bifilar detection loop.

The ideal security cable will permit detection in its continuum by ohmic methods; i.e., since the resistance per foot of the bifilar cable is known, the location of a cut can be determined.

61. Safe Industrial Grinding Wheels

More positive techniques are needed for testing the operational safety of industrial grinding wheels. Pieces flying from these wheels can cause severe injury or death. Some approaches pertained to internal bonding. Unfortunately, centrifugal action weakens the bonds. This is a very lucrative market.

62. Holographic Visualization Processes

Both the commercial and education markets have needs for more dynamic, less involved, and less expensive visualization processes. Ideally, a new invention should provide three-dimensional object renditions in full color by simple projection methods that require no "3-D glasses" or related polarizers.

This is a large order, but *holography* might provide a good answer. Basic principles were discovered some 25 years ago by the British physicist Dennis Gabor. His process could not be put to use at that time because of peculiar light requirements—the coherent, monochromatic light of the laser.

Illustrated in Fig. 3-42, holography may be regarded as lensless, three-dimensional photography. The process does not render a familiar photographic negative or print. However, if light is directed upon a holographic negative (or *hologram*), the smudgy and apparently

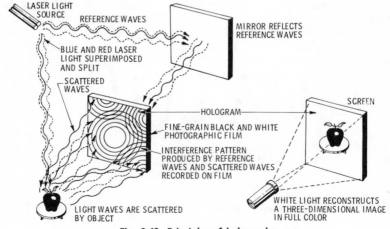

Fig. 3-42. Principles of holography.

meaningless patterns of concentric circles and parallel lines of the hologram become a "window" through which an observer sees the scene that was photographed. Moving his head from side to side, he can look through this window at different angles and change the perspective of the three-dimensional view. He can also look around the object and see what is behind it, just as if he were to examine an actual scene.

To produce a hologram effectively, light from a laser is split into two beams, one of which is directed by a mirror onto a sheet of black and white photographic film. The other beam is used to illuminate the object.

When the laser light strikes the object, it is scattered by the irregular surface and is reflected back to the film. Thus, many of the reflected light waves are jumbled and out of phase both with each other and with the light from the undisturbed beam reflected by the mirror. When the light waves from subject and mirror are reunited at the film surface, they interfere with each other in strange patterns of dark and bright areas that are recorded on the film. To cite an analogy, light waves are stored in the hologram in a manner similar to the way a musical tone is stored in a violin string; it is there, but is not released until the string is plucked or struck.

To "pluck" a hologram for release of light waves, a laser beam is passed through it. Upon hitting the hologram interference pattern, the beam is diffracted into light waves that duplicate those that were reflected from the subject. We therefore see the subject of the picture in three dimensions, apparently suspended *behind* the hologram (thus the term "window") at the same distance it was from the sheet of film.

In 1966, using principles of a photographic process originated by Nobel laureate G. Lippmann in 1908, researchers were able to obtain a multicolor image: By changing the position of the holography mirror, undisturbed laser light can be directed to the back, rather than the front, of a sheet of film. Fig. 3-42 shows this process. The beam and the scattered light reflected from the subject pass through the film in opposite directions. This results in the production of layers of interference patterns in the film emulsion. Finally, when a beam of ordinary white light is projected at the developed film, the aforementioned layers filter out all of its components, except the color of the laser beam used to illuminate the subject. So, only waves of this frequency are reflected back to the observer as a single-color, three-dimensional image.

When overlapping red and blue laser beams are used for illumination of the holographic subject, the combined beams produce even more complex layers of interference in the emulsion of the film. Now, in addition to information about intensity and phase of light reflected from the subject, we can obtain full color information, even though the hologram itself may be made on black and white film. When ordinary white light is reflected from this new hologram, two colors, red and blue, reach our eyes.

The technical problems associated with the application of holography are severe, but could be solved in time. Once invention has simplified the various processes, a most powerful and enchanting visualization technique will be available. In the entertainment field, for example, it will be possible to create dramatic scenes literally out of nothing. Elsewhere, children can be exposed to quasi-real learning experiences impossible to obtain with other methods of visual display. It is a complex problem, but it is rewarding.

63. Parachute Safety Devices

The repacking of parachutes requires great care, patience, and considerable skill. Packing by hand (Fig. 3-43) is preferred over mechanized methods because of the critical elements involved. However, especially in the military sector, there is a need for devices that can tell, at a glance, the status of a packed chute. Even if chutes are not used for a given length of time, they must be opened on packing tables and repacked. Aside from control devices for personnel-type parachutes, monitors are required for vehicular chutes. Such might

Courtesy Elsin Co.

Fig. 3-43. Packing of parachutes.

be chutes serving as air brakes for jet aircraft or as cargo chutes. Here, it is also necessary to know the status of explosive line cutters, and the condition of gores and rigging materials.

64. Improved Sculpturing Machine

A small, inexpensive sculpturing machine is needed to satisfy requirements of model shops, small-scale manufacturers, and hobbyists/artists. The machine should be able to work with soft stone, plastics, and wood. A design similar to the Wenzel sculpturing lathe (Fig. 3-44) would be suitable. Invented at the turn of the century, the lathe was able to cut two items at a time. Positional guidance

Courtesy A. Williams

Fig. 3-44. Wenzel sculpturing machine.

was furnished by a pointer that was moved over the master to be copied.

65. Storing Video and Audio in Solids

A new invention is needed to permit both visual and aural storage of program and/or data materials in solids. Similar to magnetic tape, the new design must permit rapid retrieval of specific recording increments without lengthy search procedures.

Currently, discrete *bits* of electrical data may be stored in magnetic-core memories such as shown in Fig. 3-45. Through a network of magnetic feeder circuits and interrogation loops, recording and retrieval are made possible. However, in order to store items such as audio and video, a very large and costly memory matrix would be required.

New possibilities reside in crystals. If, for example, certain Group II-VI semiconductor compounds are doped with special agents, they release a light pulse when mechanically heated or tapped. Infrared light also acts as a releasing trigger.

Fig. 3-46 shows the overall technique, involving tapping, illuminating, and heating. The electrical conductivity of the crystal jumps

Fig. 3-45. Magnetic core memory matrix.

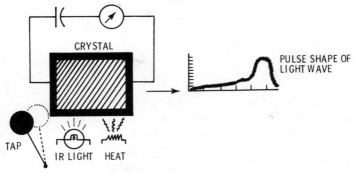

RESISTANCE-MEASURING CIRCUIT

CRYSTAL

PULSE SHAPE OF LIGHT WAVE

TAP   IR LIGHT   HEAT

Fig. 3-46. Data storage in solids.

after such stimulation, showing a negative resistance related to intensity and duration of stimulation. For additional data, see:

Y. S. Park and C. W. Litton, "Storing Light and Current in Crystals," *Electronics,* July 8, 1968, pp. 104-108.

At this time, crystals emit light only. Sodium-doped CdS crystals flash green and ZnO crystals flash blue-green when exposed to infrared radiation or tapped. Conceivably, crystals may be made to emit amplitude-modulated light. Thus, by detecting such light in a continuous mode, data would be available for readout by conventional amplifiers. Unlike storage media used in tape recording, both video and sound could be preserved in the lattice structure of a given crystal. The finalized invention would probably use little else but a small cube of solid material, its electrical contents being extracted by refined forms of the stimulation techniques mentioned above.

66. Safe High-Tension Power Switches

Line-decoupling power switches, as found in high-voltage switch yards, are likely to explode if excessive arc generation takes place during switch openings. Effective techniques are needed to suppress the electrohydraulic effect and, if possible, transfer insulating oil to an expansion vessel at very high speed. An invention for automatically bleeding condensed water from transformer and switch containers also is desired.

67. Cross-Staff for Emergency Navigation

A simple emergency-type aid is needed for navigating life boats on the high seas without a compass. The cross-staff, shown in Fig. 3-

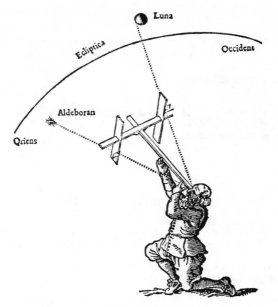

Fig. 3-47. Cross-staff for measuring angular distance.

47, was used until about 1750. The device permits determination of angular distance. A new, modern embodiment would be worthwhile.

68. Submarine Detector

No truly reliable technology is available to combat enemy submarines. However, since such vessels generate heat anomalies within the marine environment, countermeasures based on this principle could be evolved as an augmentation to sonar.

To reach heat gradients generated at great depth, a towed, deep-diving instrumentation "fish" (Fig. 3-48) might be employed. Motive

Courtesy Braincon Corp.

Fig. 3-48. Deep-diving V-fin.

power could be furnished by a helicopter or long-range blimp. Certain submarine-generated heat anomalies also can be detected by air-borne infrared photography and scanning methods. The quantum efficiency of detectors used in this area is low. Therefore, rather than a bolometer-type approach, an entirely new avenue of approach should be chosen.

69. Ecological Monitor

In ecological monitoring, scientists are interested in data pertaining to length of day, soil moisture, transparency of the sky, rainfall, and related parameters. No simple, self-contained instrumentation is available for this purpose. Fig. 3-49 suggests one approach. Here, the sensor head has been mounted on a pole. The moisture sensor is buried in the ground. To prevent vandalism at unattended field stations, the multiparameter recorder may be mounted in a camouflaged instrumentation vault. Special problems are associated with the source of operating power. Since the station is a slow-speed type, batteries that are recharged by generating solar cells may be used. Also, battery current may be used to rewind a spring drive motor. Solid-state electronics may be used, keeping dc power requirements below 200 mA maximum. Nothing similar exists.

70. Domestic Nonchemical Cleaning Utensils

Housewives would welcome a nonchemical cleaning tool for general kitchen and bathroom use. It is possible, for example, to use

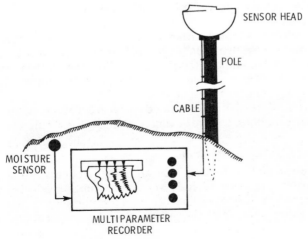

Fig. 3-49. Ecological monitoring station.

razor blades for cleaning sinks and bathtubs. If an ordinary razor blade could be combined with an electric vibrator, a superior appliance could be provided.

71. Spatial Vehicular Profile Monitor

The Victorian quotation, "See yourself as others see you," has enormous significance to pilots of surface, air, and space vehicles. For safety and other reasons, it is imperative to know the operational situation of the given craft at all times. To that end, an optical-type monitor is needed that could provide not only views of the vehicle as seen from a distance by external observers, but also "long shots" and "close-ups" as well.

The substance of the problem is shown in Fig. 3-50. Here, the viewing spot (V) has been established *behind* a manned space system. The projectors (P) provide transmitting and receiving facilities; the image is obtained in three-dimensional form.

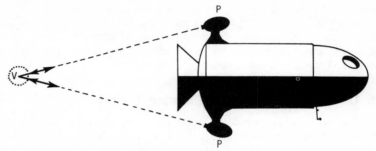

Fig. 3-50. Overall vehicular monitor.

Although an intermediate solution is possible by mounting a TV camera on an extended arm, a nonmechanical solution is desired. For example, a cesium-vapor cloud could be ejected and then scanned by laser. By allowing the cloud to function in a holographic mode via ephemeral mirrors, 3-D presentations could be achieved.

However, since the cesium-cloud approach cannot be used in atmospheric situations, a more advanced technique is required.

72. Laser Weapon Systems

For destructive purposes in warfare, large amounts of energy must be deployed and released in a very short time. Aside from agents such as nuclear fission (using uranium or plutonium) and nuclear fusion (using deuterium), weapons that can concentrate immense forces on selected targets are desired. Here, costs of general logistics and delivery should be at a minimum.

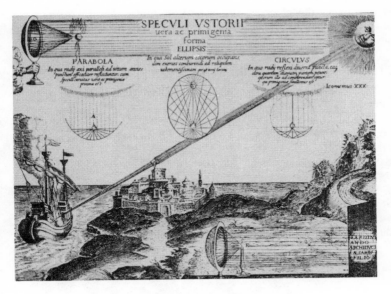

Fig. 3-51. Medieval optical weapons concepts.

Giant-pulse lasers emerge as desirable weapons in this field. These considerations are not new, however. In his book *"Ars magna lucis et umbrae,"* Athanasius Kircher projected optical armament in medieval times. Shown in Fig. 3-51, special light collectors were suggested for concentrating sunlight on ships for defensive purposes. The idea is odd but valid.

Fig. 3-52 shows an exceptionally powerful laser system with a capacity of 120 to 1200 watt-seconds. Enlarged designs can spread their energies over soft targets such as fuel farms, munitions storage areas, and the like. Unfortunately, at this time, it is not possible to increase the output of lasers by simply increasing the size of the laser rod or gas-type oscillator tube. Conventional designs pose handicaps in optical areas, atom-atom collisions, and undesirable boundary-layer interactions. The following references permit exploration of these problems:

R. W. Ditchburn, *Light* (2nd ed.; New York: John Wiley & Sons, Inc., 1963).

M. L. Stitch, "Power Output Characteristics of a Ruby Laser," *Journal of Applied Physics,* 32, 1994-99, 1961.

M. L. Stitch, "Stimulation Versus Emission in Ruby Optical Maser," in J. R. Singer, ed., *Quantum Electronics* (New York: Columbia University Press, 1961).

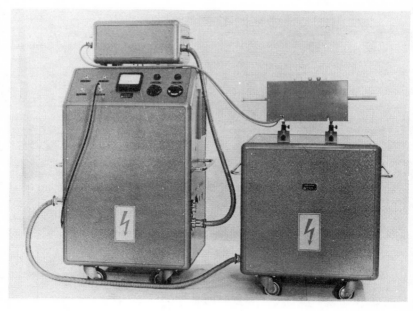

Courtesy Dr. Fruengel

Fig. 3-52. Giant-pulse laser.

73. Equipment for Investigation of Extraterrestrial and Local Phenomena

A major opportunity for invention resides in the "cosmic-heritage" area. The term embraces the inference that modern civilizations were installed by extraterrestrial agents, commonly referred to as "gods." In order to divorce these aspects from mythology and religious components, special instruments and advanced research concepts are needed.

Contemporary thought on this subject is contained in the following publications:

I. S. Shklovskkii, *"Vselennaia, Zhizn, Razum"* (*Universe, Life, Mind*), (Moscow, 1963).

I. S. Shklovskkii and C. Sagan, *Intelligent Life in the Universe* (New York: Delta, 1968).

E. Von Daeniken, *Chariots of the Gods?* (New York: G. P. Putnam's Sons, 1969).

E. Von Daeniken, *Zurück zu den Sternen: Argumente fuer das Unmoegliche,* (Duesseldorf: Econ, 1970).

J. Vallee, *Anatomy of a Phenomenon* (Chicago: Henry Regnery Co., 1965).

The substance of these arguments may be summarized as follows: The idea is projected that ancient earth was visited by galactic societies, possibly under the auspices of a missionary-type concept. It is supposed that, through extensive cross-breeding, the missionaries divorced "basic" man from his pronounced animalistic ancestry, until a reasonable likeness of man emerged. The terrestrial base of the extraterrestrials is envisioned as a common building, probably an existing structure provided by some local king, which contained the nonelectronic and electronic deep-space transmitters, stores, generators, and maintenance facilities.

Further, it is implied that the spaceship weapon systems were used for architectural purposes, and stone blocks weighing 2000 tons or more were cut from solid rock by quantum-mechanical methods. These blocks were used for the building of enormously massive structures like the Baalbek Terrace in the Anti-Lebanon Mountains, some pyramid casings, and the like. Modern-day man's desire to reach out into space is regarded as an expression of the "cosmic heritage": an attempt to return to the true cradle of our race.

A large body of scientists stands outside of these thoughts and speculations. However, their opinions should be regarded as academic safeguards and should not inhibit meaningful inquiry and progress in this extremely important field.

Not many physical artifacts are available for examination. In Bolivia, near the Peruvian border, an extremely ancient city by the name of *Tiahuanaco* is located at an elevation of two miles above sea level. Its gateway is shown in Fig. 3-53. Examinations have indicated that the city might have been pushed up from below the surface of the Pacific Ocean, since casings of sea animals could be found in its

Fig. 3-53. Gateway to mystery city of Tiahuanaco, Bolivia.

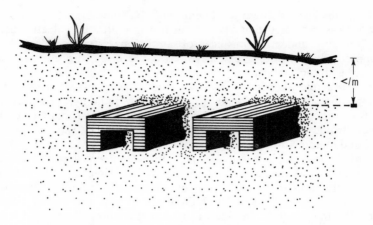

</m

Fig. 3-54. Purported "power ducts" of inverted design at Tiahuanaco, Bolivia.

remnants. Further, author Von Daeniken (see references above) discovered *inverted* ducts (Fig. 3-54) which, according to his speculations, could have served as containers for high-current power cables.

Harold T. Wilkins, in his book *Mysteries of Ancient South America,* profiled the aspects of Tiahuanaco in great depth. From an invention point of view, in an effort to solve these and other riddles, new geophysical instruments must be created in order to unlock what appear to be science-based secrets. Engineering data is included in the following references:

J. J. Jakosky, *Exploration Geophysics* (2nd ed., Newport Beach: TRIJA Publishing Co., 1960).

D. E. Parasnis, *Principles of Applied Geophysics* (New York: John Wiley & Sons, Inc., 1962).

M. B. Dobrin, *Introduction to Geophysical Prospecting* (New York: McGraw-Hill Book Company, 1960).

M. J. Aitken, *Physics and Archeology* (New York: John Wiley & Sons, Inc., 1961).

A case for cosmic-heritage inferences is made on the basis of the Piri Reis map. The map takes its name from the Turkish admiral, Piri Reis, who drew it in 1513 A.D. on the basis of maps stemming from the Alexandrine Greeks, ancient Egyptians, and/or other, still earlier civilizations.

When the map came to the attention of a retired American engineer by the name of Mallery, he noticed that the map was rather strange in certain mathematical aspects. To make a long story short, Mr. Mallery went to work with Mr. M. I. Walters of the U.S. Hydro-

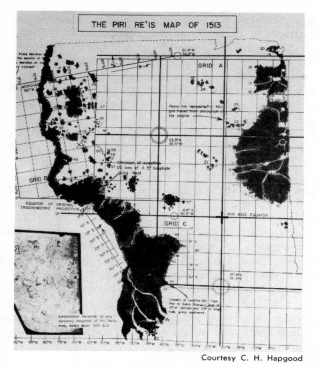

Courtesy C. H. Hapgood

Fig. 3-55. The Piri Reis Map of 1513.

graphic Office. After much patient research, a *grid* was evolved which fitted the map. The map then was corrected from that grid to modern projections (Fig. 3-55). This process resulted in a stunning discovery: The ancient map—the originals from which it had been copied dated at the latest 3000 B.C.—showed the *entire* world with great detail and accuracy! It included not only the coast lines of the Americas but also those of the entire Antarctic, including land masses in certain unexplored regions of North America.

Greatly mystified, Mr. Mallery and Mr. Walters consulted with Rev. Daniel Lineham, S.J., Director of Weston Observatory and chief seismologist for the U.S. Navy IGY explorations in Antarctica. Subsequent investigations followed. Today, all of the heretofore unknown features indicated in the Piri Reis map have been confirmed by scientific test soundings made through ice by Task Force 43.

Thus, on the basis of this evidence, the Piri Reis map emerges as an *aerial* map of the world! We must ask questions about the originators of the initial map. Who were they? Here, we are directed to the assumption that the map was photographed by superior data-gathering instruments at considerable distance from the earth. Further, production of a map of such accuracy would require the presence of mathematical aids and survey techniques we have failed to capture to this day.

In view of the above, it goes without saying that only invention can provide a set of clues to this massive complex of questions. The preceding discussion also leads us to speculation regarding orbiting observation systems of extraterrestrial origin.

On September 4, 1953, at 3:30 P.M., Mr. Charles B. Bradley of London, England, received on his television set the call letters of American TV station KLEE-TV in Houston, Texas. However, the transmission was three years late! According to Mr. Paul Huhndorf, chief engineer of the station, KLEE-TV was converted to KPRC-TV in July 1950. Where was the signal during the intervening three years? And why was it observed in England, across the Atlantic?

During 1928, Drs. Stoermer and Van Der Pol did pioneering work on the propagation of radio waves. They detected several instances in which there were not only unexplained radio echoes, but also time lags in reflected signals that reached many seconds, at times as long as one minute. When the time component was multiplied by the speed of light (also the speed of radio waves), the inference arose that the signal was reflected or rebroadcast at a distance of about 1,000,000 kilometers from earth—a considerable interplanetary distance!

It is hoped that the above will provide some points from which to start. While an orthodox approach might bring excellent results, it should also be remembered that we are likely to deal with superior technologies. If we are indeed confronting designs or concepts of extraterrestrial origin, it is advantageous to remember that "their" world was old when ours was young. Therefore, little would be gained by assuming that their means of communication, for example, would employ a relatively primitive thing like electronics. But you are the inventor. The choice is yours.

74. Photographic Scene-Preview Equipment

In flash photography, an instrument is needed to indicate effective scene illumination prior to making the actual exposure on film. Both bulb- and strobe-type flashes are too fast for retaining scene details with the unaided eye.

Fig. 3-56 suggests one approach. Here, the image of the object is directed, through the lens, onto a retaining screen that is coated with a long-persistence phosphor. The screen is arranged in a vacuum, and the image-retaining capacity of the phosphor is maintained by high-voltage excitation.

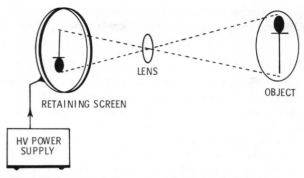

Fig. 3-56. Image-retaining scene evaluator for flash exposures.

However, because of optical and grain considerations of the phosphor used, concise tests are required to obtain acceptable image quality. Fig. 3-57 is a facsimile of an alignment test pattern evolved for optical investigations. The pattern includes resolution wedges, diagonal lines, and fine circles for determining system linearity; a bull's-eye pattern for evaluating astigmatism; and resolution patterns for checking pairing of lines.

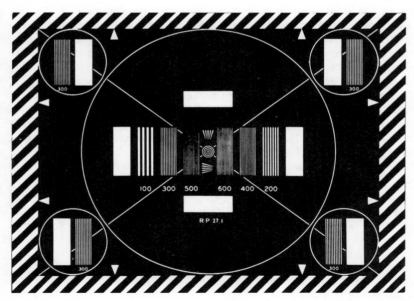

Courtesy SMPTE

Fig. 3-57. Resolution test pattern.

The final instrument should be battery-powered and, if possible, feature a zoom lens to accommodate most photographic situations.

75. Pneumatic Train

In our fight against environmental pollution, a re-examination of dated principles of transportation suggests itself. The preferred method should avail itself of space available, have no noise-generating properties, and feature a good margin of operational safety.

The compressed-air train, suggested in 1810 by George Medhurst, met some of these requirements. It was called a "pneumatic train," and power was to be furnished by air-pressure plants which forced the cars through a large tube. Unfortunately, because of psychological reasons and the unpleasantness of rapid pressure changes, the train failed to become popular. However, a new embodiment might revolve around the transportation of intercity freight. Improvements in valves and line-switching apparatus are needed to make it a reality. Concepts used in pneumatic mail systems could be adapted for this purpose.

76. Emergency Annunciator

In the operation of high-speed military and commercial aircraft, automatic monitoring of all functions is a critical neccessity. However, rather than instruments that must be watched, an acoustical alarm system would offer more benefits. A one-station emergency annunciator is shown in Fig. 3-58. The apparatus avails itself of a bank of pre-recorded tapes running continuously in loop cartridges. If an anomalous signal has been picked up by the sensor, the appropriate signal is injected into the discriminator and fed to the pilot's headset (a female voice is a better attention-getter than a male voice). It will spell out a simple message such as "Fire in engine

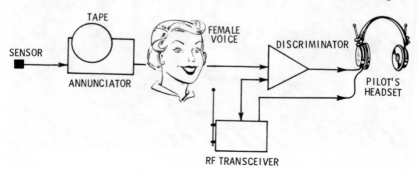

Fig. 3-58. Emergency annunciator.

two; fire in engine two." The discriminator assigns priorities to the announcements, dangerous events having top priority. Simultaneously, top-priority announcements are fed into the plane flight recorder and rf air-to-ground transmitter. Upon annunciator action, all other messages the pilot normally receives are cut off to avoid interfering cross talk.

Tapes are bulky. Improvements are needed to avoid them altogether, perhaps by using magnetic-disc systems of very small size.

77. Sheet-Metal Welder

Hobbyists would appreciate a simple, efficient machine for welding thin metal sheets together without arc-type methods or conventional principles of resistance welding. A desirable arrangement might employ, for example, a thin, small strip of current-sensitive metal. As electricity is applied, the metal melts and ensures a tight and even bond.

78. Geophysical Detectors

Throughout human history, the search for raw materials and precious metals has been an ever-refined process. Today, owing to demands made by industry and political systems, superior inventions are needed to advance explorations to new capabilities. The following examples are intended to stimulate new approaches.

The divining rod was one of the earliest instruments used in exploration activities. Author Agricola hailed its virtues in his book *De Re Metallica* (1556) and suggested a good many applications for the twig. Later, the device gave rise to modifications like the "sympathetic" pendulum. This contrivance (Fig. 3-59) contained a small specimen chamber for insertion of the metal the pendulum operator was seeking. Its gyrating motions, to the left or to the right, supposedly gave directional guidance toward treasure and the like. All of these methods belong to a group of phenomena commonly described as *occult*. Although dowsing for water, if singled out, seems to work in such far-off places as Vietnam, the art has not gained scientific respectability. An excellent summary on dowsing and its peripherals may be found in the following reference:

S. W. Tromp, *Psychical Physics* (New York: Elsevier Publishing Co., 1949).

In the United States, the official position on dowsing is expressed in a paper published by the U.S. Geological Survey:

Water Supply Paper 416 (publ. 1917).

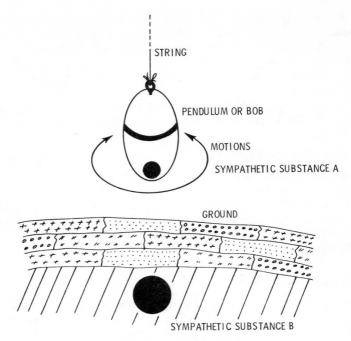

STRING

PENDULUM OR BOB

MOTIONS

SYMPATHETIC SUBSTANCE A

GROUND

SYMPATHETIC SUBSTANCE B

Fig. 3-59. "Sympathetic" pendulum.

However, while the "sympathetic" concept might appear odd, electronuclear methods can be used in a remotely similar if highly refined fashion. Fig. 3-60A illustrates the operating principle of a silver detector or "snooper." The device is capable of detecting native silver crystals or deposits as deep as three feet below ground. A ground area of about four square feet is irradiated for two minutes. As the neutron generator shoots a stream of neutrons into the earth, the silver temporarily is turned radioactive. One of its isotopes, Ag^{109}, absorbs a neutron and becomes Ag^{110}. The latter emits a gamma ray of energy 660 kiloelectron volts and has a half-life of 24 seconds. Since silver isotopes, like radioactive atoms of other elements, have a characteristic half-life, or rate of decay, and emit gamma rays at a specific energy level, the sensing system of the snooper can distinguish them from atoms of other elements, typically from elements also made radioactive in the same area by neutron bombardment.

The composite system, originally developed by the U.S. Geological Survey, is shown attached to a vehicle in Fig. 3-60B. Technically, the power for producing neutrons comes from a front-mounted gasoline generator that runs a simple particle accelerator. A beam of

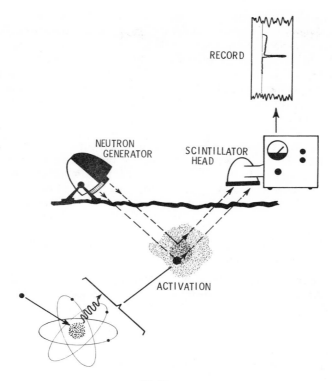

RECORD

NEUTRON GENERATOR

SCINTILLATOR HEAD

ACTIVATION

(A) Diagram

(B) Equipment

Fig. 3-60. Silver detector using neutron activation.

deuterium (heavy hydrogen) particles emitted by the accelerator is directed against deuterium absorbed in a titanium target. As the deuterium particles collide, they release neutrons that are channeled into a beam which covers a four-square-foot area of ground.

This technique is not overly complex and lends itself to searches for other minerals, typically gold, copper, vanadium, and aluminum. To make instrumentation of this type available to the average prospector, its price tag of about $30,000 must be brought down to a more modest figure, say, $2,000. This simplified machine might shoot low-energy X-rays at the ground, causing silver deposits or other minerals to fluoresce. The fluorescence may then be measured with a scintillation counter or other device of similar sensitivity. A technique of this type is, of course, vastly superior to "beeping" treasure detectors that use induction loops. Magnetometers cannot be used here, since the materials involved are nonmagnetic, nor do they cause pronounced geomagnetic anomalies.

The following reference should aid independent efforts in this exceptionally promising field:

> F. E. Senftle and A. F. Hoyte, "Mineral Exploration and Soil Analysis Using in Situ Neutron Activation," *Nuclear Instruments and Methods* 42 (1966), 93-103 (North-Holland Publishing Co.)

79. Building/Mountain X-Ray Techniques

In conjunction with the preceding discussion, new inventions are needed to "X-ray" the interior structures of ancient buildings or monuments, and to obtain diagnostic data on hidden caves, underground faults, and the like. The cost of the desired instrumentation should be within the means of schools, museums, and private research groups.

Currently, two methods are available for this purpose: film and spark chamber. The active agents in both cases are cosmic rays.

In the film-type method, ray-sensitive photographic emulsions are placed in a cave beneath the structure to be investigated. If the latter contains hollow sections or cavities of interest, the film will "see" more exposure. If, however, a given structure is strictly solid, uniform exposure will be obtained. Cavities permit higher particle speeds than do intervening materials such as rock. By placing different sheets of film at different locations in the detector station, it is possible (theoretically) to obtain an idea of anomalies in the overhead burden.

A more advanced technique involves the use of spark chambers and associated monitoring equipment. The method was tried experi-

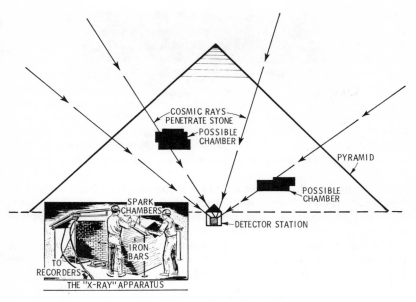

Fig. 3-61. X-raying a pyramid.

Courtesy Lawrence Radiation Laboratory

Fig. 3-62. X-ray detection gear for Chephren Pyramid.

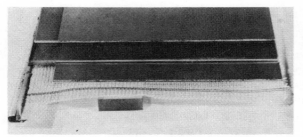

Fig. 3-63. Details of spark chamber for Chephren Pyramid.

mentally in an effort to locate hidden chambers in the Pyramid of Chephren, Egypt. The overall approach is illustrated in Fig. 3-61. The detection equipment minus iron bars is shown in the photograph of Fig. 3-62.

Natural cosmic rays, some powerful enough to pass through the pyramid (in this case), come from all directions in the sky. Spark chambers, installed at the bottom of the pyramid, register the direction from which the rays come. If there is a large number of rays from a given angle, a hollow feature, such as a hidden burial chamber, is indicated in that direction.

Fig. 3-63 shows the spark chamber used. Two copper-clad plates are positioned a fraction of an inch apart. As the cosmic-ray particle passes through the plates, a high-voltage spark follows its trail be-

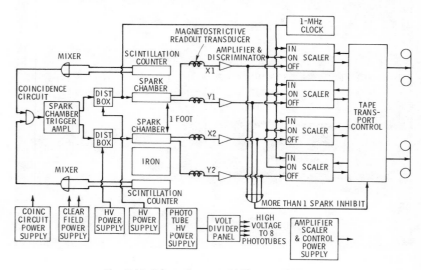

Fig. 3-64. Schematic, pyramid X-ray equipment.

tween them. This allows recording of the location of the impact of the particle, either by photographing the spark or, better yet, by sensing and registering its position directly on magnetic recording tape.

As shown in Fig. 3-64, two spark chambers are mounted in a tray, one a foot below the other. Thus, the relative location of the passage of a cosmic ray through each one shows the angle of the trajectory of the ray.

Fig. 3-61 shows a five-foot deep stack of iron bars below the spark chambers. The reason is an important one: If the image of, say, a 300-foot distant cavity is fuzzy, the cosmic ray having traversed its "empty" volume should have enough energy left to go through the iron bars. Therefore, *scintillators* have been arranged beneath the stack. These units, when struck by a cosmic-ray particle, trigger the spark chambers to record the event. The spark chambers do not respond to cosmic rays of less energy.

Unfortunately, the equipment was not able to detect discrete chambers in the Pyramid of Chephren. According to a comment made by Dr. Amr Gohed of Cairo University, results obtained are ". . . scientifically impossible . . . reference points, which should emerge at each recording, cannot be seen." Others agree. Either the interior of the pyramid is criss-crossed by unknown shafts and chambers or, to quote Dr. Gohed again, "We are dealing with a mystery which is beyond orthodox explanations—call it occultism, curse of the pharaohs, magic, or whatever you wish!"

Needless to say, this is a tremendous challenge to inventors.

80. Dissipator for Static Electricity

A generally applicable material is needed for dissipating static electricity, and unpleasant shock, from carpets and other items made of plastic fibers. This would be an excellent seller in the automotive field, too.

81. Refrigerator Oil Separator

Manufacturers of refrigerating equipment would appreciate a separator for removing oil from freon or ammonia. Since oil is finely vaporized in the compressor, mere traces of oil lead to insulating problems.

82. Blind-Spot Eliminator for Cars

Optical or other inventions are desired to overcome the dangerous "blind spot" in the left rear section of automobiles. Rear-vision mirrors normally do not cover this area.

83. Induction-Type Iron

A cordless iron would find an excellent market in both home and commercial fields. Since no lasting, high-powered batteries are available, the *induction iron* shown in Fig. 3-65 might be an intermediate solution. Current is furnished by an rf generator to the induction coil arranged beneath the ironing board. The iron itself has a built-in induction loop in its base plate. Thus, the latter constitutes the secondary winding of a simple transformer circuit operated in a short-circuit mode.

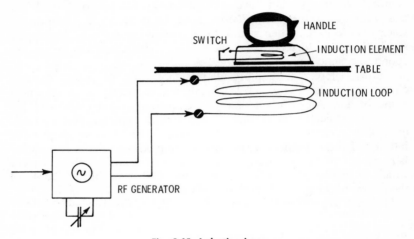

Fig. 3-65. Induction iron.

84. Noninterfering Windshield Wiper

There should be a wide market for a new type of automotive windshield wiper that will not move in front of the driver's face. An entirely new method for keeping water from the windshield is desired.

85. Soap Initials

In order to retain identification throughout use of their product, soap manufacturers would welcome a simple process for injecting initials or other designs into cakes of soap. Suitable letters could be formed from a soap base different in color but identical in wear characteristics to the main soap body. The invention should avoid excessive smear during use, thus maintaining advertising values.

86. Substitute for Tire Chains

A substitute for tire chains is needed. Oval wheels have been tried, but they keep the speed low and vibration high.

87. Soft-Tire Alarm

A simple alarm system is needed to warn drivers of soft tires.

88. Separator for Disabled Aircraft

In military air operations, it is imperative that combat-disabled planes, such as burning aircraft, do not damage planes or facilities located near runways. Desirably, means should be available to *isolate* impaired planes fully in case of trouble. Special walls, to be erected by hydraulic means, have been suggested. However, because of the great expense and random factors involved, such proposals could not be realized. This would be a very rewarding market.

89. Hydraulic Transmission for Locomotives

For economy and other reasons, a superior hydraulic transmission is needed for diesel-powered locomotives. Contemporary diesel-electric arrangements are expensive to install and to maintain. The desired transmission should have zero slip, stay cool, provide reliable torque on grades, and have a self-braking feature. The problem of a reliable traction system is associated with the above.

90. Automotive Cylinder Monitor

A simple instrument is desired to show proper functioning of all cylinders of an automobile engine under all load conditions.

91. Ash Compressor for Cars

For automotive use, there is need for an ash try that will suppress the glow of cigarette butts. This might be accomplished with a simple ash compressor.

92. Summer Windows for Motorists

To restrict the entrance of insects, a summer screen in place of car windows would find an excellent market.

93. Neon Fog Lights

Red neon lamps penetrate fog much better than incandescent types do. A simple neon tail light, combined with an inexpensive and reliable solid-state power supply, would be acquired readily by many car owners.

94. Machine Tool Speed Control

For superior speed control of such machine tools as lathes and drill presses, a variable dc drive system is desired. Besides eliminating the need for shifting pulleys and belts, the drive would permit speed variations during work. Aside from a torque-controlled dc drive, a small, efficient *hydraulic* transmission may also be inserted between a standard prime mover and the machine.

95. Nuclear Deep-Space Antenna Projector

In deep-space communications, effective data transmission and reception is contingent upon antenna systems of superior quality. Very high gain is one of the most desirable features. With transceiving distances between interplanetary expeditions and earth likely to increase, attention should be directed toward advanced antenna structures and principles.

At the right of Fig. 3-66 is an umbrella-type S-band antenna used during a recent moon mission. Assembled on a tripod, this device has a collapsible feature that saves vehicular storage space and permits rapid erection on site. However, where very great distances are involved, such simple designs are not likely to suffice.

Fig. 3-67 suggests an advanced approach. Instead of using any kind of wire in the construction of antenna elements, a radioactive-particle projector is indicated. Since given particles conduct radio-frequency currents, a particle beam envelope can be used for transmission and reception purposes. The improved invention will permit operations in any ambient medium, i.e., vacuum or atmospheric gases. The particle projector must be safe to handle and use, preferably emitting a radiation dosage of less than 0.001 mr/hour.

For a discussion of antenna gain, consult:

Reference Data for Radio Engineers (5th ed., Indianapolis: Howard W. Sams & Co., Inc, 1968) pp. 25-41 to 25-43.

For a discussion of space communication, see:

Ibid., Chapter 34.

Fig. 3-66. S-band antenna at moon base.

96. Barnacle Suppressor for Ships

Both military and commercial operators of ships would welcome an effective invention for preventing formation of barnacles. These shell-bearing sea animals not only change the hydrodynamic profile of a ship, but can also interfere with critical gear such as sonar. Barnacle-repelling paints contain poison, but lose their effectiveness very rapidly. The animals cannot attach themselves to vessels moving at speeds greater than 3 knots.

Barnacles, like other sea animals, fear electricity. A given invention might start with investigation of physiological behavior in electric fields. If, for example, definite susceptance to rf can be observed, a conductive plastic covering or paint could be used as an rf-conveyor below the water line of the ship. A general discussion on the application of electricity to sea animals is given in the following book:

L. G. Lawrence, *Electronics in Oceanography* (Indianapolis: Howard W. Sams & Co., Inc., 1967) p. 238 ff.

97. Geophysical Shock-Wave Generators

In the seismic phase of exploration geophysics, a strong impulse wave may be generated by explosives. However, since placing of chemicals involves digging of shot holes and rather continuous "hustling" (carrying) of seismophones from one location to the next, more economical and rapid generation of shock waves is desired.

A good, functional approach has been developed by Sinclair Research, Inc. The method, called "Dinoseis," employs a confined explosion of propane and oxygen to pump energy through shock waves into the ground. Basic to the Dinoseis is a seismic pulse generator mounted on a converted 7½-yard earth mover. The gas exploder is mounted to a hydraulic jacking mechanism that replaces the scraper pan of the earth mover.

In operation, the detonation chamber is pressed on the ground at selected shot intervals, the gas explosion being radio-triggered by an observer in the recording truck. Three or more units may be used at the same time, with explosions set off simultaneously. Shock waves

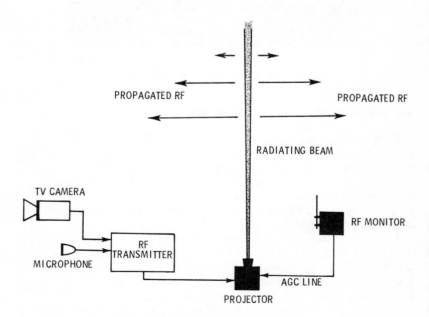

Fig. 3-67. Particle-type antenna projector.

thus generated penetrate the ground and return signals for sensing by a spread of seismophones and recording on a digital field system.

Inside the pulse generator is a floating plate, free to move with respect to the 20,000-pound top weight, including the weight of the machine itself. Inertia of the top weight is very great. So, when the gas mixture is detonated, the lighter 500-pound bottom plate responds to the shock wave, transferring about 100,000 foot-pounds of energy to the surface of the earth in 2 milliseconds. The rapid transfer of energy gives a sharp pulse with a wide range of effective frequencies, resulting in sharp reflections from buried strata.

However, overall, the field is wide open for new inventions. It is possible, for example, to generate a very powerful, explosion-like event by discharging high voltage under water. Little else but water and an adequate power supply are required to produce results with this *electrohydraulic effect*. Specific problems reside in areas of pulse application to the soil and rapid recycling of storage capacitors. Basic references may be found in the following article:

L. G. Lawrence, "Electrohydraulic Effect," *Electronics World,* May 1969, p. 44 ff.

98. Radiofrequency Catalyst for Insect Control

One of the greatest boons to our society would be insect control by nonchemical or semichemical methods lacking polluting side effects. It is episodic pollution that captures public attention. The recent deaths of thousands of sheep in the vicinity of a military testing ground are an example. However, within the context of the problem, perhaps more important are long-range cumulative effects produced by chlorinated organic pesticides such as aldrin, DDT, dieldrin, methoxychlor, and others. The dangerous philosophy prevails that if one cannot see, feel, smell, or taste a pollutant, it does not exist. Nuclear hazards, for example, were frequently overlooked until obvious cases of radiation sickness were reported. Here it took inventors to provide advanced instrumentation that replaced the old spinthariscope in this critical field.

Fossil records indicate that insects are the oldest inhabitants on earth. Ninty percent of all animals are insects. Silk, cochineal, honey, and lac are insect products. A few hundred of the 500,000 species known are harmful. Selective insect control in agriculture and peripheral fields is important to effective plant growth and yield. Although the intrinsic nature of insects is not fully understood, it appears that electronics could emerge as a controlling agent.

Insects are susceptible to ultrasonic waves and electromagnetic fields. If dense energy trains are directed at a specimen, internal heating effects tend to occur, and the animal might boil and explode. However, if gains are evaluated against costs, the fact emerges that the use of electronic methods alone is too expensive for practical applications.

By contrast, chemicals are cheap. This invites us to consider combinations of low-cost electronics with inexpensive chemicals, possibly in a catalytic sense. *Enzymes,* for example, act as catalysts in insect metabolic processes. The same holds true of water, since many life-sustaining compounds react only in solutions. Thus, electronic pulse trains might be directed at insect populations that have eaten catalyst-type chemicals responsive to relatively weak electromagnetic stimulation. Here, the catalyst would give rise to abnormal *endergonic* states resulting in death. A *normal* endergonic event is the combination of carbon dioxide and water to form sugar within a living system.

A basic approach is shown in Fig. 3-68. The endergonic-type organic catalyst is applied to crops in the form of a spray. After the catalyst is taken up by insects 12 or more hours later, radiofrequency energy triggers it into action inside the body of the insect. Lethality will result from metabolistic anomalies in populations thus radiated. The ideal spray would function as a catalyst only in small animal life forms. It would have no effect on plants *per se,* nor would it be harmful if not irradiated by rf.

The following references provide a basic introduction:

A. Hollaender, ed., *Radiation Biology,* Vols. I, II, and III (New York: McGraw-Hill Book Company, 1956).

G. G. Simpson, *Life: An Introduction to Biology* (New York: Harcourt, Brace and Co., 1957).

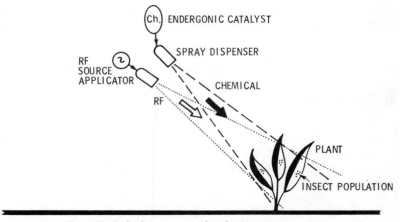

Fig. 3-68. Radio-frequency catalyst for insect suppression.

99. Plant Electroculture

In conjunction with environmental pollution, attention is directed to the nitrate pollution of agricultural fertilizers. New methods, preferably electrical, are desired to stimulate plant growth and yield.

Historically, attempts to increase the growth of plants date back to the 18th century. Dr. Mambray at Edinburgh, Scotland, apparently was the first to conduct experiments in 1746. Major experiments were conducted by Dr. S. Lemstroem in Finland in 1903. Excellent results were obtained.

Dr. Lemstroem, a professor of physics at Helsingfors, came to the belief that very rapid growth of vegetation in polar regions during the short arctic summer was to be ascribed to special electrical conditions of the atmosphere in these high latitudes. He duplicated these assumed conditions by increasing the atmospheric current, which normally passes from the air to the plant, by the use of antenna-type wires placed above the crop. An electrostatic Wimshurst machine was used for his purpose. Results were given in Dr. Lemstroem's book, *Electricity and Agriculture and Horticulture,* (London, 1904).

Lemstroem's work and results gave rise to experiments on an international scale, as is reflected in the following literature:

F. Basty, *Nouvaux Essais d'Electroculture* (Paris: C. Amat, 1910).

V. H. Blackman, *et al,* "The Effect of an Electric Current of Very Low Intensity on the Rate of Growth of the Coleoptile of Barley," Proc. Roy. Soc. B., 95, 214-28, 1923.

K. Stern, *Elektrophysiologie der Pflanzen* (Heidelberg-Berlin: J. Springer, 1924).

The various methods are known under the combining term "electroculture." Fig. 3-69 shows a typical electroculture system. The dc

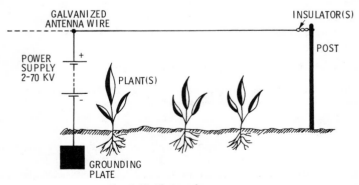

Fig. 3-69. Electroculture system.

exciting voltages are determined by the height of the feeding antenna, but usually are between 2000 and 70,000 volts. The following observations have been made:

1. The true percentage of yield increase for a good field is 45 percent maximum.
2. The better a field has been plowed, the greater is the yield obtained through electroculture. If the soil is too lean, no pronounced increase in yield can be observed.
3. Some plants do not respond to treatment unless watered. If they are watered, the yield can be extremely high. Peas, carrots, and cabbage are typical.
4. Electroculture treatment is detrimental to many, probably all, plants if conducted on warm, sunny days.
5. Overhead antenna wires should be arranged high enough to permit convenient plowing.

In the vicinity of the plant, current densities radiated by overhead discharge systems range approximately from 10^{-12} to 10^{-11} A/cm². *Natural* electrical current densities peak out between 10^{-16} to 10^{-15} A/cm². Although these current densities might appear to be extremely low, note that electroculture provides currents that are about 1000 to 10,000 times higher than those given by nature. Electrometers may be used in the field to establish proper current levels. Higher drive voltages are needed to compensate for increased antenna height.

The domain of electroculture is wide open for new inventions because experiments are not complete. Specific needs for improvements reside in the area of in-field high-voltage generation and application.

As a closing note, it is of interest to realize that plant electroculture, in spite of its excellent promise, was relegated to dormancy by the advent of inexpensive nitrate fertilizers. Here we have a typical case in which a once-fabulous invention in chemistry commenced to dig its own grave (and perhaps ours, too!).

100. Psychogalvanic Effects in Living Plants

Although living plants normally are not regarded as more than edibles or raw material for construction purposes, new inventions and experiments are needed to determine their *sentient* properties.

Early in 1966, polygraph expert Cleve Backster of New York created the background for an entirely new science by accidental discovery. Employing the same kind of polygraph (lie detector) that is used to test emotional stimulation in human subjects, Backster found that plant specimens register fear, apprehension, pleasure, and relief. Further, by using simple electronic methods, it was found that plants react not only to overt threats to their state of well-being, but even more stunningly, to the intentions and feelings of other living creatures, animal as well as human, with which they are closely associated.

It could be shown, for example, that simple house plants, such as the Dracaena Massangeana or philodendron, register apprehension when a dog goes past them, react violently when live shrimp are dumped into boiling water, and apparently receive signals from dying cells in the drying blood of an accidentally cut finger. Plants even appear to respond to distress signals over a considerable distance.

The immense importance of what is now referred to as the "Backster effect" need not be underlined. The phenomenon offers entirely new possibilities to inventors and research scientists, typically in areas of communications heretofore closed to us.

The overall aspects of the "Backster effect" were profiled in the following publications:

C. Backster, "Evidence of a Primary Perception in Plant Life," *International Journal of Parapsychology,* Vol. 10:4, Winter 1968, pp. 329-48.

Anon., "ESP: More Science, Less Mysticism," *Medical World News,* Vol. 10:12, March 21, 1969, pp. 20-21.

L. G. Lawrence, "Electronics and the Living Plant," *Electronics World,* October 1969, pp. 25-28.

L. G. Lawrence, "Electronics and Parapsychology," *Electronics World,* April 1970, pp. 27-29.

The effect continues to be verified both here and abroad. Its action cannot be blocked by Faraday screens, screen cages, lead-lined containers, or other shielding structures positioned between the plant and external test objects. Therefore, the phenomenon cannot be added to the inventory of electromagnetic domains established by classical physics.

How does the effect come about? Unfortunately, there are no concise answers at this time. The field is too new. However, the effect has a *psychogalvanic* character and can be verified by instrumentation batteries such as that shown in Fig. 3-70. A typical test system is composed of a variable Wheatstone bridge, a dc amplifier, a Faraday cage for the plant specimen, and a graphic recorder for collecting data in a permanent manner.

In operation, the cage-contained plant is connected by a simple, leaf-attached clamp electrode to the Wheatstone bridge, and the readout system is energized. Then, by *mentally* projecting physical

Courtesy Electro-Physics Co.

Fig. 3-70. Testing psychogalvanic effect in plants.

harm against the plant, response curves may be elicited. However, if the threat is not followed up by physical action (like burning, for example), the plant tends to adjust to these "idle threats" and ceases to respond. Response profiles are not uniform, changing from one specimen to the next. No responses and/or *delayed* reactions occur in many cases, all of which are imperfectly understood.

As a guide to inventors, the following equipment systems, approaches, and scientific considerations are offered as practical aids:

1. To avoid industrial and domestic electrical interference, electronics-oriented experiments with plants should be conducted in shielded enclosures such as metallic greenhouses. If, however, such a structure is too costly, an inexpensive wooden design may be used. Here, interference can be attenuated by attaching metallic screen wire to the walls, bottom, and ceiling of the greenhouse. Note that light must be permitted to enter, since it is vital to photosynthesis. If a given Faraday-type enclosure is totally opaque, artificial illumination must be provided. If lamps of the fluorescent type are employed, light must enter through a meshed metal screen securely grounded to a water pipe. For information on lamps designed explicitly for living plants, contact:

Fig. 3-71. Investigating psychogalvanic effect of desert plants with in-field electronics.

Commercial Engineering Department
Lighting Division
Sylvania Electric Products Inc.
Salem, Massachusetts

Request bulletins 0-262 ("Gro-Lux" fluorescent lamp), 0-285 and 0-286. The visible light spectrum required by plants for successful chlorophyll synthesis is between 4000 and 4800 angstroms, effective values peaking out at about 4500 angstroms.

2. For field operations involving testing of plants, psychogalvanic equipment must be housed in rugged, shielded containers (Fig. 3-71). This requirement is especially stringent in those cases in which susceptibility tests are performed adjacent to power lines, populated areas, or automotive service stations generating voltage transients. Electrically, plants may be regarded as *organic* semiconductors and, regardless of size, have orthodox antenna functions.

3. For defendable experimental results, new, virgin plants may be used. Fresh cultures may be started conveniently from packaged growth kits such as "Punch 'N Grow" provided by Northrup, King & Co., Minneapolis, Minnesota. The kit contains

seeds and soil material in a plastic container. Its cover may be punched with a blunt tool, and water may be added for seed activation.

4. In conjunction with item 3, seeds may be stimulated and raised into the seedling stage under the influence of weak radiofrequency fields. Apparatus for this purpose should operate below a wavelength of 10 meters. For information on rf stimulation of seeds, perusal of the following research paper is suggested:

> P. A. Ark, "Application of High-Frequency Electrostatic Fields in Agriculture," *Quarterly Review of Biology,* 1940, 15:172-191.

In the high-power stimulation phase, rf energy is applied for fractions of a second. According to the scheme shown in Fig. 3-72, seeds are contained in a plastic sack or bowl and placed between application plates connected to the rf generator tank circuit. This process is most critical and should not be attempted without background studies.

In the low-power stimulation phase, rf currents in the microampere or milliampere range are applied to a *planted* seed specimen. Apparatus is shown in Fig. 3-73. The blower-cooled rf generator to the left feeds energy into an interval switch at the bottom of the quasi-Faraday cage. A nutrient feeder for seedlings, seen at the upper right of the illustration, furnishes weak auxin-type solutions or other growth hormones.

Because of its radio-frequency character, equipment of this type must *always* be operated within an electromagnetically shielded facility such as those outlined in item 1.

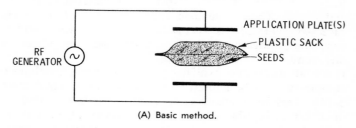

(A) Basic method.

STIMULATED (SHORT TIME BASE)

CONTROL

STIMULATED (LONG TIME BASE)

(B) Yield patterns.

Fig. 3-72. Stimulation of seeds by rf methods.

Courtesy Electro-Physics Co.

Fig. 3-73. Composite rf stimulator for seeds and seedlings.

As an experimental adjunct to rf methods, seeds also can be stimulated successfully by ultrasonic processes. A research paper on the ultrasonic treatment of corn seeds is available from:

U.S. Department of Agriculture
Washington, D.C. 20250

A piezoelectric-type ultrasonic generator is shown in Fig. 3-74. Operating at a frequency of 165 kHz, the instrument requires no rectifier and drives the crystal at a power of 17 watts.

In the ultrasonic mode, seeds to be treated are placed in a sieve-type metal container and placed in a water bath agitated by ultrasonic waves. Frequencies up to 900 kHz have been used. According to reports, ultrasonics are yield-stimulating; i.e., seeds treated by this method germinate better than controls and render more produce. However, within the context of the current problem, the method is suggested primarily in order to grow virgin plants fast for experimental purposes. As in the case of rf methods, no information is available on how rf or ultrasonically stimulated plant properties differ in the expression of psychogalvanic phenomena in connection with the "Backster effect."

(A) Equipment.

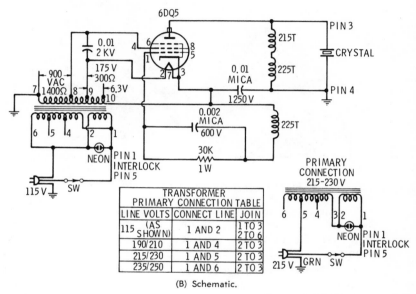

TRANSFORMER PRIMARY CONNECTION TABLE		
LINE VOLTS	CONNECT LINE	JOIN
115 (AS SHOWN)	1 AND 2	1 TO 3 / 2 TO 6
190/210	1 AND 4	2 TO 3
215/230	1 AND 5	2 TO 3
235/250	1 AND 6	2 TO 3

(B) Schematic.

Courtesy L & R Manufacturing Co.

Fig. 3-74. "Minisonic" ultrasonic system.

5. To commercialize an invention in this area, given equipment should be small and functional. Being a "discovery tool" for others (including students, engineers, gardeners, hobbyists, etc.), new psychogalvanic circuitry should be developed to display plant reactions with the least possible expense.

Fig. 3-75 shows one approach. A current-controlled oscillator (Fig. 3-75A) forces a weak excitation current through a plant leaf via a clamp electrode (Fig. 3-75B). Variations in the conductance properties of the plant, which may be regarded as an organic semiconductor, will "steer" current-sensitive oscillator stage Q1. Thus, overall action is similar to that of Fig. 3-75C —a Wheatstone-type detector principle.

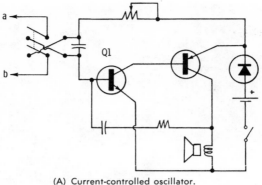

(A) Current-controlled oscillator.

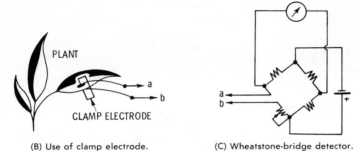

(B) Use of clamp electrode.   (C) Wheatstone-bridge detector.

Fig. 3-75. Basic acoustical plant response detector.

Convenient packaging of the above circuit is illustrated in Fig. 3-76. Arranged on a tripod, the plant response detector might feature an elevated clamp electrode mounted on an extension inserted into the instrument proper. Thus, plant structures can be reached and observed as desired. During operation, it is imperative that the electrode fixture does not move. Movements or bending of leaves inserted into the clamp electrode produce undesirable strain-gauge effects and impair the quality of psychogalvanic data.

In conjunction with this and related equipment, the inventor should realize that plant reactions, if any, cannot be predicted in advance. Fire (injury by acts of burning) is the most powerful stimulant a plant can experience. It was the mental projection of this very threat that led Mr. Backster to the discovery of the effect named after him. By contrast, the act of cutting can bring somewhat delayed reactions. Electrical plant behavior triggered by *direct* application of force produces well-known responses, typically as investigated in a specimen like the

Fig. 3-76. Acoustical plant response detector in use.

Mimosa Pudica. Data and directions for experiments may be found in the following reference:

J. Bures, *et al, Electrophysiological Methods in Biological Research* (New York: Academic Press Inc., 1967).

The function of organic semiconductors, if considered alone, is discussed in good depth in:

F. Gutmann, *Organic Semiconductors* (New York: John Wiley & Sons, Inc., 1967).

The behavior of living systems exposed to stimulus and excitation is given in the book:

G. Ungar, *Excitation* (Springfield, Ill.: Charles C Thomas, Publisher, 1963).

An excellent description of electron transport systems in plants has been given by Walter D. Donner, Jr. of the University of Pennsylvania. His contribution may be found in:

W. A. Jensen and L. G. Kavaljian (ed.), *Plant Biology Today: Advances and Challenges* (2nd ed., Belmont, Calif.: Wadsworth Publishing Company, Inc., 1967).

All things considered, the psychogalvanic effect in plants is one of the very best vehicles for dramatic new discoveries and instruments available to inventors today. The effects of these phenomena on science and industry no doubt will be considerable and of immense value to society at large. To that end, it is hoped that definitions of the various aspects involved point ways from which to start.

101. Prevention of Aircraft Hijacking

An effective device is needed to preclude hijacking of aircraft. Chloroform has been suggested for injection into passenger compartments to put would-be hijackers to sleep. However, owing to side effects, such a method cannot be used, and a more advanced invention is sought.

102. Vacuum-Type Eraser

To eliminate "gumming-up" of typewriters by an accumulation of eraser residues, there is need for an invention that combines erasing with removal of residues. A vacuum-type implement might be suitable for this purpose.

103. Earthquake Forecasting

A functional earthquake forecasting system would be a mandatory component of every nation's geophysical observatories around the world. Its enormous benefits, especially in humanitarian and material perspectives, need not be underlined. Unfortunately, the *triggering* mechanism of quakes continues to escape precise definition. However, some progress is being made in peripheral areas.

The destructive power of earthquakes is well known. In investigations, the first point fixed by seismologists is the depth of the underground center of the earthquake, the *focus,* or *centrum.* Seismographic measurements are compared for this purpose. Although it was once supposed that earthquakes never originated below a depth of 40 miles, recent discoveries have evidenced focus depths far below the lithosphere of the earth crust, i.e., depths down to 435 miles.

One of the major causes of earthquakes appears to be faulting: either movement along an old fault or development of a new one. Volcanic activity at a depth and surface eruptions produce quakes that are small or of moderate strength. Even less intense are seismic events caused by landslides and subsidence in caverns and mines. The latter might be an effect of earthquakes rather than their cause.

The Benioff strain seismometer (Fig. 3-77) continues to be a successful instrument for measuring accumulated stress in the earth.

Fig. 3-77. Benioff strain seismometer.

The transducer end, featuring a low-loss capacitive pickup, is shown in the foreground. As the capacitive area changes as a result of stress on the long quartz tube, the output frequency of an rf circuit changes as well. Photoelectric and similar pickup modes may be used alternately. Only the transducer end is movable; the other end of the tube is rigidly attached to a pier. The Benioff system is not responsive to electrical phenomena associated with earth movements.

A more advantageous technique resides in the detection of *piezomagnetic* properties or *magnetostrictive* effects of rock under stress. A *piezoelectric* effect, if considered alone, may come about in certain layered and/or faulted overburdens undergoing stress or strain. However, if subsurface stress manifests itself as a change in the susceptibility and remnant magnetization of rock, the local geomagnetic field should change as a result. Stress precedes an actual earthquake. Thus, by determining the rate of stress increase, conclusions may be reached on the build-up of events leading up to an actual earthquake.

Fig. 3-78 profiles field instrumentation. A magnetometer and laser interferometer are shown. The method is complementary. In its basic form, the interferometer compares optical wavelength with a standard of length, typically by means of interference fringes. This principle may be modified to measure a Doppler shift in optical frequency if ground movements occur as a result of piezodynamic effects intially observed by the recording magnetometer. For stronger movements or shocks, a conventional seismometer may serve as a monitor, possibly in a standby mode.

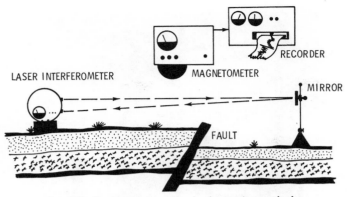

Fig. 3-78. Experimental earthquake forecasting methods.

Geophysical-type magnetometers required for this work must feature long-term stability and noise-cancelling provisions. A sensitivity of at least 0.1 gamma (γ), or 10^{-6} oersted, is mandatory. Several magnetometer designs meet this requirement electronically, but have certain mechanical disadvantages. Sensitivity to seismic vibrations and slight misalignment restrict their in-field use. The rubidium- and helium-vapor types are superior, but are extremely costly and therefore difficult to acquire for unsponsored exploratory experiments.

A good magnetometer that could be refined is the cathode-ray type, shown in developmental form in Fig. 3-79. Here, slight motions of the electron beam are detected by a photoelectric cell, and the cell output is fed to a readout. If such an instrument combines positioning feedback, superior sensitivity up to 0.001 gamma appears feasible. The electron beam normally remains stationary on the CRT screen. Deflections are evoked only by external magnetic fields, such

Fig. 3-79. Developmental magnetometer.

Fig. 3-80. Magnetometer calibration chamber.

as the magnetic field of the earth, acting upon it. A further advantage of this type of magnetometer is its natural ability to indicate magnetic polarities without excessive data processing.

However, magnetometers require accurate calibration facilities regardless of make. The Marshall test and calibration chamber (Fig. 3-80) uses precisely controlled magnetic fields for its purpose. In this particular arrangement, Helmholtz coils are fastened to an equipment deck. Metered currents are furnished by the auxiliary unit to the right of the illustration. Magnetometer heads may be placed within the Helmholtz coils directly.

The above is but a fair sample of what can be done in this challenging field. Note, however, that data can be difficult to interpret in many ways. It is necessary, for example, to be cognizant of diurnal and other geomagnetic variations associated with our planet. Not only does the magnetic field change throughout the day and the year, as shown in Fig. 3-81, but it may be influenced by local industrial interferences as well. Thus, for earthquake forecasting to be effective by virtue of magnetic instruments, given interferences must be suppressed in order to obtain true data only.

The field is wide open and in desparate need of new inventions. The following references are some of the best available internationally:

S. Breiner, *The Piezomagnetic Effect in Seismically Active Areas* (Final Report, November 1967, Division of Mathe-

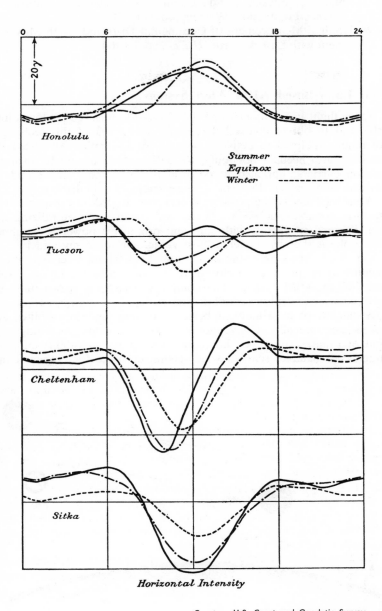

Horizontal Intensity

Fig. 3-81. Diurnal variations of horizontal geomagnetic intensity.

matical, Physical and Engineering Sciences, National Science Foundation, Washington, D.C. 20550).

W. Honig, "Measurement of Continental Drift and Earth Movement with Lasers," *Proc. IEEE,* April 1964.

104. Laser-Based ABM Alert System

Space flight has demonstrated the ability to put large payloads into orbit and to obtain controlled re-entry. Principles of this technique are readily adapted to delivery systems for H-bombs. Advanced inventions are needed not only to detect and deactivate orbiting weapon carriers, but also to destroy, in space, any rocket and its payload entering the domestic defense sector in a ballistic trajectory.

An assessment of the problem is complex. In order to inspect satellites for possible nuclear warheads and commence neutralizing action, manned spacecraft with sufficient propulsive power is required. The probe is thereby rendered most vulnerable to enemy attack.

Fig. 3-82 shows an approximate approach. Here, laser-type interrogation concepts are employed to provide high-resolution target data for an ABM system. The composite weapons carrier must be detected *prior* to deployment of its decoys.

By means of an alignment beam, two interrogation satellites are linked electronically. Conventional rf transmissions, including any kind of radar, cannot be chosen because of the possibility of jamming. Each satellite scans a wide section of space for targets. If an

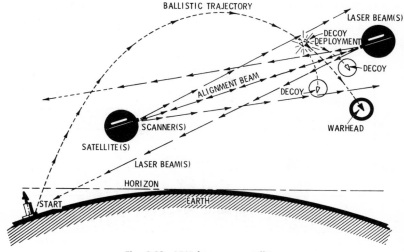

Fig. 3-82. ABM laser-scan satellites.

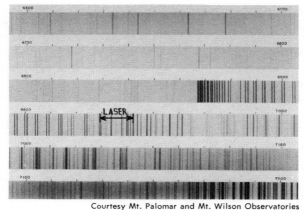

Courtesy Mt. Palomar and Mt. Wilson Observatories

Fig. 3-83. Infrared spectrum of sun.

optical return is received, target triangulation is triggered. Information in the form of coordinates is made available, in optical spectrums, to ground-based defense systems.

While this concept is superior to conventional countermeasure techniques, several complex solutions to peripheral and intrinsic problems are required.

An infrared-type laser transceiving system, operating in the customary giant-pulse mode between 6929 and 6943 angstroms, is susceptible to optical interference by sunlight (Fig. 3-83). Therefore, the equipment is required to use an intermediate data carrier when facing the sun. This may be accomplished by using a *redar* (not radar!) system such as that shown in Fig. 3-84. This concept com-

Courtesy Prirecon Co.

Fig. 3-84. Microwave/laser redar diplexer.

bines a radar-type diplexer with a high-power ruby laser (Al_2O_3, Cr_{3+}). Thus, the laser beam may be fed past keep-alive radar components by conventional "plumbing" and may be emitted in a straight line of sight from the *horn* of the conventional parabolic reflector. Optical action recommences as soon as the sun interference is attenuated by positional tilt of the satellite.

Efficient action of the laser is possible by selecting a nonluminous "window" in the sun spectrum. Windows have small spectral widths, but can be used if the quantum-mechanical efficiency of the photo-multiplier tube in the laser transceiver is high at the range of interest. Consult respective S-numbers for guidance.

Aside from the complexities involved, an invention in this area requires adequate funding for basic experiments. However, market potentials are excellent.

105. Prevention of Explosive Suit Decompression

In space environments, techniques are needed to prevent injury to personnel caused by explosive decompression of equipment and/or impact with objects. Overall effects are dramatized in Fig. 3-85. A beaker of fluid, here shown in a simulated situation in a high-

Courtesy U.S. Air Force

Fig. 3-85. Explosive decompression under simulated conditions.

Fig. 3-86. Tangent galvanometer of Cornell University (1890).

altitude chamber, is brought to rapid expansion due to decompression. The event would be fatal if encountered in real life.

106. Research Magnetometers

Problems attached to the measurement of magnetism have never been simple ones. There is a need for magnetometers that feature superior operational characteristics, yet are low in cost. A sensitivity of 1 gamma (γ) or better is desired.

Dipping compass needles were the earliest magnetometer designs. Evolved in Sweden, instruments of this type saw extensive use in searches for iron ore. Later, based on experiences with Helmholtz coils and tangent galvanometers (Fig. 3-86), better magnetic materials evolved for use in magnetometer systems. This, in time, gave rise to flux-gate instruments, search-coil designs, and finally, a small family of superb but very expensive magnetometers such as the rubidium- and helium-vapor types.

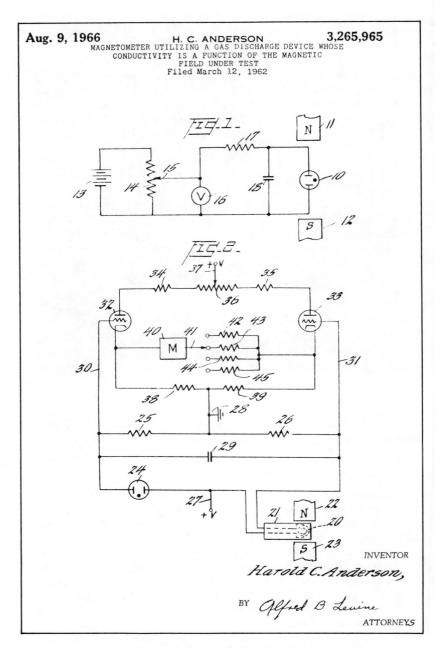

MAGNETOMETER UTILIZING A GAS DISCHARGE DEVICE WHOSE
CONDUCTIVITY IS A FUNCTION OF THE MAGNETIC
FIELD UNDER TEST
Filed March 12, 1962

INVENTOR

Harold C. Anderson,

BY *Alfred B. Levine*

ATTORNEYS

Fig. 3-87. Patent pertaining to gas-discharge magnetometer system.

Since these instruments are of considerable importance to military, commercial, and scientific vendors, innovative inquiry should be directed into this area.

One of the simplest magnetometer designs known is based on tangent-galvanometer principles (Fig. 3-86). Here, a magnetic body is sensitive to flux existing external to it. Resultant movements, either spin or dip, may be observed directly or with magnifying optical aids. Although such a magnetometer would require no significant external power for its operation, sensitivity to vibration and shock delimits its features and use.

One of the least involved electronics-type magnetometers was invented by H. C. Anderson (U.S. Patent 3,265,965). Schematically shown in Fig. 3-87, the apparatus avails itself of a gaseous discharge device whose degree of ionization, or electrical conductivity, is a function of the magnetic field acting upon it. An amplifying bridge circuit *compares* the conductivity of a second discharge device (No. 24 in the schematic) with that of device No. 20 positioned in a magnetic field. Very low magnetic domains, such as geomagnetic anomalies, cannot be detected with this device because of inadequate sensitivity, but improvements could bring this feature about.

107. Improved Flash Boiler

New methods are needed to advance the efficiency of automotive flash boilers. The general acceptance of steam cars hinges on safe, reliable boiler designs having very fast start-up times. An approximate state of the art is shown in the advertisement for Barton flash

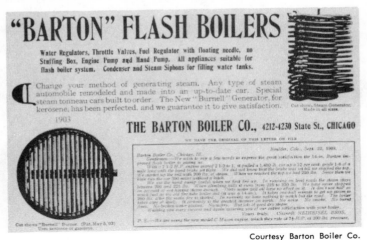

Courtesy Barton Boiler Co.

Fig. 3-88. Advertisement for early automotive flash boiler (1903).

boilers, Fig. 3-88. The term "flash" is derived from the fact that water injected into the steam-generator tubing *flashes* into steam. The unit is not given to explosions as "teakettle" boilers are. Major improvements in burner designs, effective boiler surface, and water injection techniques are needed.

108. Deep-Space Laser Engines

In the United States today, sponsored programs of research and development in the field of electric propulsion are aimed at manned and unmanned space vehicles, primarily for interplanetary missions. Unfortunately, current approaches are awkward at best. The power yields obtained with ion-type and related electric space engines are low. Superior inventions are needed to advance the state of the art. *Genuine* pioneering efforts are invited.

Fig. 3-89 profiles the *status quo*. Here, high-power chemical engines are envisioned for providing vehicular lift from the planetary surface and limited maneuvers in space. Special shuttle craft is projected for (1) transportation of personnel to and from orbiting stations, and (2) carrying of fuel and equipment to vehicles under con-

Courtesy Douglas Aircraft Co., Inc.

Fig. 3-89. Assembly of space ship in space.

struction. Because of shortcomings attached to chemical deep-space propulsion, large carriers cannot be assembled at terrestrial sites.

Of the advanced propulsion concepts suggested, quantum-mechanical systems such as lasers invite special consideration. Lasers have the inherent ability to provide immense shock waves in plasma and brute force in the form of *light pressure.*

According to calculations by Nobel laureate Arthur L. Schawlow, the following facts on laser-generated light pressure have emerged: Peak powers of 500 million watts have been reported in laser beams with a cross section of less than 1 cm². Assuming that the beam *intensity* is about 10^9 watts per cm², the intensity of the corresponding electric field is nearly 10^6 volts per centimeter in the unfocused beam. A good, conventional lens with a focal length of one centimeter could focus the spot to a thousandth of a centimeter in diameter. Thus, the beam intensity in this focal spot would be a million billion (10^{15}) watts per square centimeter, and the *optical-frequency* electric field would be about one billion (10^9) volts per centimeter. In such a focused beam, light or radiation pressure would be over 15,000,000 pounds per square inch.

Through the use of laser batteries that fire in continuous sequence into a suitable pressure-transmitting agent, immense propulsion energies would be available. Unfortunately, when materials for transferring agents are considered, the power product is higher than that which binds the outer electrons in most atoms. It would cause severe disruptions even in transparent substances and produce a nonpropulsive "drill-through" effect in conventional materials, including hard substances such as natural and industrial diamonds.

At this time, it is not known how high-energy laser beams react on a semicritical nuclear mass or plasma. However, if fission could be suppressed in a counteracting mode, nuclear substrates might conceivably transfer components of mechanical force, pressure *per se,* to a vehicular frame without the possibility of a nuclear detonation.

The action of laser beams of *different* frequency needs to be examined, especially in plasma-generated shock waves. Fig. 3-90 shows an approximate arrangement. Here, a spectrum analyzer looks at an 800 MHz difference between two coherent laser beams. Effective display is made possible by the exceptional sweep width of 2 GHz of the equipment (a Hewlett-Packard Model 851B/8551B analyzer is shown). However, for purposes of discussion, single-beam action will be described.

An experimental laser engine concept is shown in Fig. 3-91. Fig. 3-92 illustrates a developmental unit standing adjacent to a conventional liquid-fuel rocket motor.

In operation, as shown in Fig. 3-91, a giant-pulse laser unit injects its focused beam into a shock chamber. Plasma is generated in front

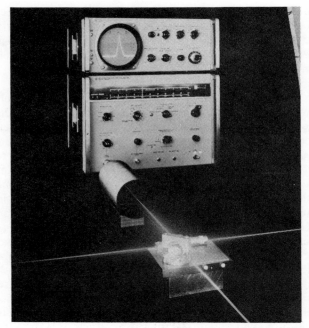

Fig. 3-90. Laser-beam mixing and display on spectrum analyzer.

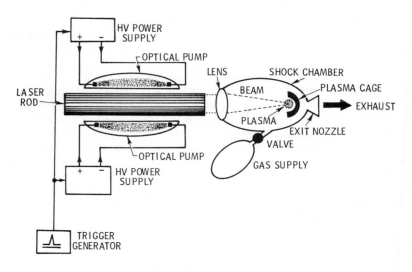

Fig. 3-91. Laser engine.

Courtesy Prirecon Co.

Fig. 3-92. Developmental laser engine at right, liquid-fuel rocket motor at left.

of a plasma-retaining cage, with inert gas or common air serving as a plasma-generating medium.

Gas may be maintained at an intitial pressure of 1000 to 2000 psi. Here, electron densities as high as 10^{20} electrons/cm^3 will be produced with an optical input energy of 2 to 3 joules, or watt-seconds.

The laser-formed gaseous plasma has a powerful explosive character. Strong blast waves, or gasdynamic shocks, and a "fireball" may be observed experimentally.

Although plasmas seem to form without delay, Kerr-cell photography shows a specific growth and extinction profile. As shown in the

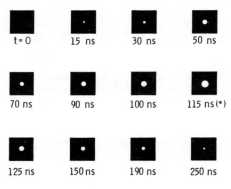

(*) MAXIMUM

Fig. 3-93. Laser-developed growth of plasma in argon maintained at 2000 psi.

drawings of Fig. 3-93, plasma growth reaches a maximum after a time lapse of approximately 115 ns, with rapid extinction setting in after 250 ns. The illustration pertains to plasma in an argon atmosphere maintained at 2000 psi. Higher pressures *stretch* the decay time. In argon gas maintained at 20,000 psi, for example, a residual plasma component remains photographically discernible after 850 nanoseconds.

Problems associated with effective laser engine development reside in impulse-generation and shock-transmission areas. For shock to be effective as a prime mover, pulse sequences must be projected in a jointed and/or overlapping mode. Further, no plasma-forming medium such as gas should be required for engine operation. Problems in vehicular logistics can be mitigated, perhaps, by using metal vapors derived from solids, including available materials such as rocks.

Laser engines bear exceptional promise for long-term space missions. Because of the absence of nuclear-type radiation, preference over atomic power plants is implied. The following references are suggested for studies:

D. W. Gregg and S. J. Thomas, "Plasma Temperature Generated by Focused Laser Giant Pulses," *J. Appl. Phys.,* 38, March 15, 1967, pp. 1729-1731.

O. M. Friedrich, *et al, Investigation of Strong Blast Waves and the Dynamics of Laser-Induced Plasmas in High-Pressure Gases* (AIAA Paper No. 67-696, AIAA Electric Propulsion and Plasmadynamics Conference, Colo., 1967).

C. E. Bell and J. A. Landt, "Laser-Induced High-Pressure Shock Waves in Water," *Appl. Phys. Ltrs.,* 10, January 15, 1967, pp. 46-48.

109. Blood-Powered Fuel Cell for Heart Pacer

In the medical area, new inventions are needed for providing electrical power to aids such as heart pacers. Currently, battery replacement entails surgery. The life span of even the best of batteries is limited, typically three to five years. An overall discussion is given in the following reference:

J. T. Prentice, "Electronic Implants," *Electronics World,* June 1968, pp. 46 ff.

An advanced approach is the blood-to-electricity converter shown in Fig. 3-94. Suggested by J. Fishman and J. Henry, the device takes the form of a fuel cell. Selective *catalysis* is the key. Although blood carries many compounds, each electrode catalyzes *only one*

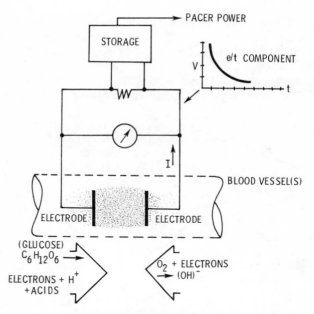

Fig. 3-94. Blood-to-electricity converter.

reaction of *only one* compound. The electrodes are coated with gold-palladium alloy; the anode coat is 55 percent gold; the cathode coat is 85 percent.

When inserted into blood, the anode breaks down glucose (a sugar) into hydrogen ions (an acid) and electrons. At the cathode, the blood oxygen takes up electrons to form hydroxyl ions. In a load circuit across the electrodes, electrons flow from the region of the glucose reaction through the load to the oxygen electrode. The following article provides more insight:

Anon., "Blood Power," *Electronics,* July 20, 1970, pp. 45-46.

Typical power outputs of blood-operated fuel cells are low, about 20 μW. Staggering-type arrangements are needed to increase output. A power product of 0.5 watt or more is desired. Here, safeguards against electrolysis are required.

110. Medical Shrapnel Detector

Special metal detectors are needed to pinpoint very small shrapnel fragments in the human body. The method should provide superior resolution, but should not subject patients to dangers of radiation sickness caused by repeated X-ray studies.

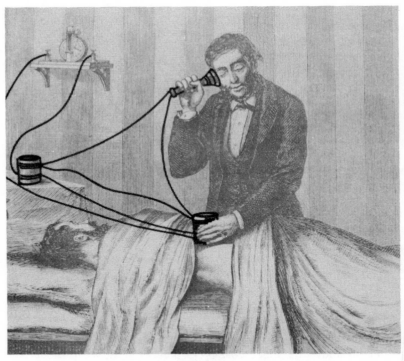

Fig. 3-95. Dr. Hughes' medium-resolution shrapnel detector.

Induction-type methods were suggested in 1881 by Dr. Hughes (Fig. 3-95). Note that large inductors were used. This technique has not changed in principle. However, since most military shell fragments are magnetic, an ultraminiature magnetometer-type sensor might be employed.

111. Noiseless Naval Propulsion Systems

Noiseless propulsion systems are needed for antisubmarine warfare (ASW). Currently, if a destroyer wishes to probe the marine depths in silence, diesels or turbines must be shut off, and the propeller(s) must be stopped. This makes the ship a "sitting duck," since the submarine is listening for target returns as well. Various methods have been suggested, but fully tried only in selected cases.

Fig. 3-96 shows the principle of "silent" oars. Here, oars are dipped gently into the water by a steam-powered frame. However, this method of propulsion is both dated and extremely awkward to use in modern naval warfare.

Fig. 3-96. "Silent" oars powered by engine-driven structure.

Fig. 3-97A shows the *Flettner* rotor system, evolved in the early 1920's by Anton Flettner in Germany. Spinning cylinders of considerable height replaced the sails on his "rotorship." Use was made of the *Magnus* principle, which states that, if a ball or cylinder spins in the path of a current of liquid or gas, there is a force on the rotating object at right angles to the direction of the current.

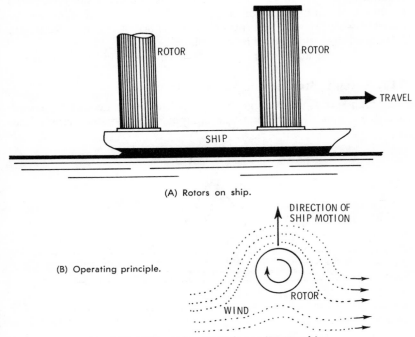

(A) Rotors on ship.

(B) Operating principle.

Fig. 3-97. Flettner rotor system for marine propulsion.

Typically, the Magnus principle accounts for the fact that a ball, flying through air with a spin, curves off to one side. We see this in soccer games.

In Fig. 3-97B, a motor-driven cylinder is seen from above. The wind streams from the left to the right. Assuming the cylinder is given a clockwise spin by the ship engine, the difference in pressure on the sides turning with and against the wind will move the cylinder in the direction of the arrow. Reversing the direction of spin reverses the course of the cylinder. Both cylinders on the ship normally spin in one direction. The ship will turn if the cylinders spin in opposite directions; thus no rudder is needed to turn.

A Flettner-type naval vessel, if equipped with a shock-mounted engine for driving the cylinders, has noiseless features. However, the main disadvantage is that the cylinders represent dominant radar targets. Perhaps a new invention could eliminate this, giving "soft" or no radar returns.

Fig. 3-98 shows a ship with a rotary wind-velocity converter. In essence, the arrangement is based on a large windmill or propeller driven by the wind. A drive shaft and gear box transfer the resultant power to a conventional propeller below the water line. A heavy keel is required for stability, since the ship is essentially a sail-type vessel requiring air in motion for its operation. Provided that noises associated with the gear box, propeller, bearings, etc., can be kept very low, naval operation would be possible.

Note that the above principles can be embodied in action toys.

112. Circuit Breaker for "Live" Structures

A high-speed electromagnetic circuit breaker is needed to disconnect both domestic and industrial equipment from power supply lines

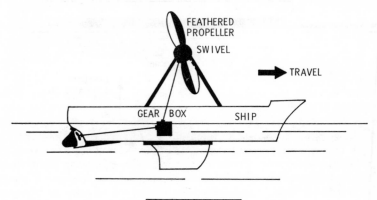

Fig. 3-98. Rotary wind-velocity converter.

in case of "hot" shorts. Here, unless the equipment is grounded by three-wire cord, metal parts will carry live electricity with consequent danger of shock. The breaker must respond to both high and low leakage. Experience has shown that even currents in the milli-ampere range cause heart failure and death.

113. Multipurpose Transducer

Multipurpose transducers are needed for indication and recording of physical forces. In the electrohydraulic field, for example, shock-wave pressure generated in water by spark discharge embodies components of hydraulic and electrical power. The desired transducer should be suited to operate with *two* discrete sets of forces. Current methods deal with one effect only.

Fig. 3-99 suggests an approximate approach. The device shown is an ammeter design that emerged prior to the turn of the century.

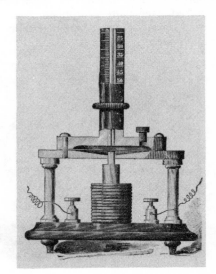

Fig. 3-99. Electrohydraulic ammeter.

Current through the coil at the bottom pulls a movable core down, activating a diaphragm and causing the mercury in the tube above the diaphragm to fall. In a modern version of this dated design, pressure increments could be applied to the diaphragm while the coil retains its original purpose, measurement of current. Photosensitive paper could be moved across the scaling tube, the latter being illuminated by short-term (1 millisecond) stroboscopic flashes. A simple fiducial marker could be injected to identify hydraulic or electrical events.

114. Camera Exposure Monitor

In both professional and amateur photography, more convenient methods of exposure determination are desired. A simple exposure-meter *monitor,* connected by cable to the internal meter of a camera, might be mounted to eyeglasses to permit more convenient reading by persons using such appliances.

115. Spring-Driven DC Chopper

A small, spring-driven chopper for dc amplifiers is needed. Mechanical choppers have quasi-zero resistance when on, and infinite resistance when off. These features cannot yet be emulated by solid-state or related chopping devices. Ideally, the spring-type drive system could activate a dc generator as well. Power derived from such a source for field instruments equipped with low-drain IC's would be preferred over battery-supplied potentials. Batteries may operate in a stand-by mode, yet be smaller than they would need to be for full-time service.

Courtesy J. I. Case Company

Fig. 3-100. Two-horse treadmill.

116. Treadmill for Power Generation

In many underdeveloped countries, electrical power is not conveniently available for medical and school use. Thus, the native population cannot receive the benefits of X-rays, nor are children in isolated villages privileged to be taught, by actual experiments, the basics of modern technology. It is in these and related areas that animal-powered generating systems could be reinvented.

Fig. 3-100 shows the highly successful Case two-horse treadmill power system advertised in 1886. Because of obvious advantages, treadmills continued to be manufactured in the United States long after the advent of steam power.

The operation of a circular saw is shown. If an electric generator is to be driven by this method, dc would be the best choice. If the generator is equipped with a fast-acting voltage regulator, power can be fed into a solid-state converter and ac can be obtained. In Europe, Africa, etc., the converter output power would be 110/220 volts ac at 50 Hz.

117. Battery-Powered AC Motor

A battery-powered ac motor is needed for small phonographs and tape recorders. Dc motors now in use generate too much electrical noise to permit successful transfer of music or speech to conventional ac machines. The desired motor might use only one converter transistor for its purpose, the motor field providing essential transformer action. A feedback loop needs to be added to trigger and sustain circuit oscillation. Battery-powered ac motors are found in motion-picture type recorders or portable TV recorders only.

118. Plastic Electronics Cabinets

After having completed assembly of an electronic instrument, most hobbyists desire to install it in a rugged, professional-looking enclosure. Unfortunately, cabinets and boxes available today do not meet these needs adequately. Cheap packaging is chosen, even though the given instrument has superior professional qualities. Innovation is needed to bridge the gap between extremes. *Plastic* enclosures and cabinets would find an excellent market, especially if carefully styled to have a professional look. Punch-out holes could be featured on front and rear panels. Requirements of shielding can be met by furnishing self-adhesive foil with the cabinets.

119. Sprinkler Guard

Automatic sprinkler units used in fire protection of industrial facilities often release water when there is no fire. At the other extreme, because of corrosion or local electrolysis of active parts, water may be released only after damaging temperatures have been reached. A new invention is needed to release water only when a preset temperature (within a ± 10-percent tolerance) has been attained.

120. Brain-Wave Weapon System

Different concepts of warfare that disable rather than kill or destroy are sought. The easiest way, as suggested by "humanitarians," would be to have antagonistic political leaders fight each other with clubs in a pitch-dark room. The one who emerges alive the next morning is the winner; all the people in his nation are winners.

Of course, it would be impossible to carry this out in practice. An invention that provides brain-wave control could, however, accomplish good deeds in behalf of innocent populations who otherwise would have to face each other in physical warfare.

The power of electrical oscillations in the atmospheric and ionospheric systems of the earth varies geographically. The frequency of oscillations may be raised by increasing, in an artificial manner, the number of lightning strokes. Here, in view of the fact that electrical activity in the human brain is concentrated at certain frequencies, suppressant psychological states could be achieved.

Experiments have been performed in the use of a flickering light to force the *alpha* rhythm of the human brain into unnatural synchronization with the light source—visual stimulation leading to electrical stimulation. Also, in experiments discussed by Dr. Norbert Wiener, a sheet of tin plate was suspended from a ceiling and connected to a generator working at 10 hertz. When a field strength of one or two volts per centimeter was made to oscillate at the alpha-rhythm frequency of the brain, very unpleasant sensations were noted by human subjects.

The scheme is farfetched, but enhanced low-frequency electrical oscillations in the earth-ionosphere cavity invite definite considerations relating to nonlethal weapons systems. Background investigations are aligned with biological research. The following references permit insight:

W. Adey, *et al,* "Impedance Measurements in Brain Tissue of Animals Using Microvolt Signals," *Exp. Neurol.,* 5:47-66, 1962.

G. Amantea, "Ueber experimentelle beim Versuchstier infolge afferenter Reize erzeugte Epilepsie," *Pflueg. Arch. Ges. Physiol.,* 188:287-297, 1921.

A. Arvanitaki, *Les variations graduées de la polarisation des systèmes excitables* (Paris: Hermann Cie., 1938).

L. J. Viernstein and R. G. Grossman, "Neural Discharge Patterns in the Transmission of Sensory Information," *Information Theory,* 20:262-269, 1961.

J. Bures, *et al, "Electrophysiological Methods in Biological Research* (New York: Academic Press Inc., 1967).

General Information Concerning Patents

Appendix A consists of excerpts from *General Information Concerning Patents*. This publication, of special interest to those who desire to patent an invention, is designed to acquaint inventors with the procedures and policies of the Patent Office. It contains a discussion of what can be patented, how and by whom a patent application should be made, what drawings are required, and other items of interest. It is not intended to supply sufficient information for the inventor to file a patent himself; in almost all cases this should be done by a patent attorney or a patent agent who has full knowledge of the legal procedures involved.

General Information Concerning Patents may be purchased for 15 cents from:

> Superintendent of Documents
> U. S. Government Printing Office
> Washington, D.C. 20402

Contents

For sale by the Superintendent of Documents, U.S. Government Printing Office
Washington, D.C., 20402 - Price 15 cents

which was enacted July 19, 1952, and which came into effect January 1, 1953. This law is reprinted in a pamphlet entitled *Patent Laws*, which is sold by the Superintendent of Documents, U.S. Government Printing Office, Washington 25, D.C.

The patent law specifies the subject matter for which a patent may be obtained and the conditions for patentability. The law establishes the Patent Office for administering the law relating to the granting of patents, and contains various other provisions relating to patents.

WHAT CAN BE PATENTED

The patent law specifies the general field of subject matter that can be patented, and the conditions under which a patent may be obtained.

In the language of the statute, any person who "invents or discovers any new and useful process, machine, manufacture, or composition of matter, or any new and useful improvements thereof, may obtain a patent," subject to the conditions and requirements of the law. By the word "process" is meant a process or method, and new processes, primarily industrial or technical processes, may be patented. The term "machine" used in the statute needs no explanation. The term "manufacture" refers to articles which are made, and includes all manufactured articles. The term "composition of matter" relates to chemical compositions and may include mixtures of ingredients as well as new chemical compounds. These classes of subject matter taken together include practically everything which is made by man and the processes for making them.

The Atomic Energy Act of 1954 excludes the patenting of inventions useful solely in the utilization of special nuclear material or atomic energy for atomic weapons.

The statute specifies that the subject matter must be "useful." The term "useful" in this connection refers to the condition that the subject matter has a useful purpose and also includes operativeness, that is, a machine which will not operate to perform the intended purpose would not be called useful. Alleged inventions of perpetual motion machines are refused patents.

Interpretations of the statute by the courts have defined the limits of the field of subject matter which can be patented, thus it has been held that methods of doing business and printed matter cannot be patented. In the case of mixtures of ingredients, such as medicines, a patent cannot be granted unless there is more to the mixture than the effect of its compounds. (So-called patent medicines are not patented; the phrase "patent medicine" in this connection does not have the meaning that the medicine is patented.) It is often said that a patent cannot be obtained upon a mere idea or suggestion. The patent is granted upon the new machine, manufacture, etc., as has been said, and not upon the idea or suggestion of the new machine. As will be stated later, a complete description of the actual machine or other subject matter sought to be patented is required.

PUBLICATIONS OF THE PATENT OFFICE

Patents.—The specification and accompanying drawings of all patents are published on the day they are granted, and printed copies are sold to the public by the Patent Office. Over 3,000,000 patents have been issued.

Printed copies of any patent identified by its patent number, may be purchased from the Patent Office at a cost of 25 cents each, postage free, except design patents which are 10 cents each.

Future patents classified in subclasses containing subject matter of interest may be obtained, as they issue, by prepayment of a deposit and a service charge. For the cost of such subscription service, a separate inquiry should be sent to the Patent Office.

Official Gazette of the United States Patent Office.—The Official Gazette of the United States Patent Office is the official journal relating to patents and trademarks. It has been published weekly since January 1872 (replacing the old Patent Office Reports), and is now issued each Tuesday, simultaneously with the weekly issue of the patents. It contains a claim and a selected figure of the drawings of each patent granted on that day; decisions in patent and trademark cases rendered by the courts and the Patent Office; notices of patent and trademark suits; indexes of patents and patentees; list of patents available for license or sale; and much general information such as orders, notices, changes in rules, changes in classification, etc. The Official Gazette is sold, by annual subscription and in single copies, by the Superintendent of Documents, Washington, D.C.

Beginning with July 1952, the illustrations and claims of the patents are arranged in order according to the Patent Office classification of subject matter. New patents relating to any particular field can thus be located more readily. Copies of the Official Gazette may be found in public libraries of larger cities.

Index of Patents.—This annual index to the Official Gazette contains an alphabetical index of the names of patentees and a list identifying the subject matter of the patents granted during the calendar year. Copies are sold by the Superintendent of Documents. At present it is issued in two volumes, one for patents and one for trademarks.

Decisions of the Commissioner of Patents.—There is issued annually a volume republishing the decisions which have been published weekly in the Official Gazette. Copies are sold by the Superintendent of Documents.

Manual of Classification.—This is a loose-leaf book containing a list of all the classes and subclasses of inventions in the Patent Office classification of patents, a subject matter index, and other information relating to classification. Substitute pages are issued from time to time. The manual and subscriptions for the substitute pages are sold by the Superintendent of Documents.

Classification Bulletins.—The various changes and advances in classification made from time to time are collected and published in bulletins which give these changes and the definitions of new and revised classes and subclasses. These bulletins are sold by the Superintendent of Documents and the Patent Office.

Patent Laws.—A compilation of the patent laws in force is issued, and revised editions are published from time to time. Copies may be purchased from the Superintendent of Documents.

Trademark Laws.—A compilation of the trademark laws in force is issued, and revised editions are published, from time to time. Copies are sold by the Superintendent of Documents.

Rules of Practice of the United States Patent Office in Patent Cases.—This publication contains the rules governing the procedures in the Patent Office which have been adopted by the Commissioner under the authority of the patent statutes and approved by the Secretary of Commerce, and supplementary material including forms and relevant sections of the patent law. Copies may be purchased from the Superintendent of Documents.

Trademark Rules of Practice of the United States Patent Office.—This publication contains the rules governing the procedure in the Patent Office in trademark matters. Copies may be purchased from the Superintendent of Documents.

General Information Concerning Patents.—This pamphlet is designed for the layman and contains a large amount of general information concerning the granting of patents expressed in non-technical language. Single copies are distributed by the Commissioner free on request.

General Information Concerning Trademarks.—This pamphlet serves the same purpose with reference to trademarks as the preceding does concerning patents.

Patents and Inventions, an Information Aid for Inventors.—The purpose of this pamphlet is to help inventors in deciding whether to apply for patents, in obtaining patent protection and in promoting their inventions. Copies may be purchased from the Superintendent of Documents.

Roster of Attorneys and Agents Registered to Practice Before the United States Patent Office.—This list of registered attorneys and agents, arranged alphabetically and by States and cities is sold by the Superintendent of Documents.

Patent Attorneys and Agents Available to Represent Inventors Before the United States Patent Office.—This pamphlet lists only those patent attorneys and agents registered to prepare and prosecute patent applications before the Patent Office, who are available to represent individual inventors and companies. Attorneys and agents who are not available to represent individual inventors and companies because they are employed by a corporate employer or by the U.S. Government, are not listed.

arranged set of patents. These libraries and their locations are: Albany, N.Y., University of State of New York; Atlanta, Ga., Georgia Tech Library; Boston, Mass., Public Library; Buffalo, N.Y., Buffalo and Erie County Public Library; Chicago, Ill., Public Library; Cincinnati, Ohio, Public Library; Cleveland, Ohio, Public Library; Columbus, Ohio, Ohio State University Library; Detroit, Mich., Public Library; Kansas City, Mo., Linda Hall Library; Los Angeles, Calif., Public Library; Madison, Wis., State Historical Society of Wisconsin; Milwaukee, Wis., Public Library; Newark, N.J., Public Library; New York, N.Y., Public Library; Philadelphia, Pa., Franklin Institute; Pittsburgh, Pa., Carnegie Library; Providence, R.I., Public Library; St. Louis, Mo., Public Library; Stillwater, Okla., Oklahoma Agricultural and Mechanical College; Toledo, Ohio, Public Library.

The Patent Office has also prepared on microfilm lists of the numbers of the patents issued in each of its subclasses, and many libraries have purchased copies of these lists. In libraries which have the lists and a copy of the Manual of Classification, and also a set of patent copies or the Official Gazette, it will be unnecessary for the searcher to communicate with the Patent Office before commencing his search, as he can learn from the Manual of Classification the subclasses which his search should include, then identify the numbers of the patents in these subclasses from the microfilm lists, and examine the patent copies so identified, or the disclosures of these patents in the Official Gazette volumes.

ATTORNEYS AND AGENTS

The preparation of an application for patent and the conducting of the proceedings in the Patent Office to obtain the patent is an undertaking requiring knowledge of patent law and Patent Office practice as well as knowledge of the scientific or technical matters involved in the particular invention.

The inventor may prepare his own application and file it in the Patent Office and conduct the proceedings himself, but unless he is familiar with these matters or studies them in detail, he may get into considerable difficulty. While a patent may be obtained in many cases by persons not skilled in this work, there would be no assurance that the patent obtained would adequately protect the particular invention.

Most inventors employ the services of persons known as patent attorneys or patent agents. The statute gives the Patent Office the power to make rules and regulations governing the recognition of patent attorneys and agents to practice before the Patent Office, and persons who are not recognized by the Patent Office for this practice are not permitted by law to represent inventors. The Patent Office maintains a register of attorneys and agents. To be admitted to this register, a person must comply with the regulations prescribed by the Office, which now require a showing that the person is of good moral character and of good repute and that he has the legal and scientific and technical quali-

by patent attorneys and agents for their professional services are not subject to regulation by the Patent Office. Definite evidence of overcharging may afford basis for Patent Office action, but the Office rarely intervenes in disputes concerning fees.

WHO MAY APPLY FOR A PATENT

According to the statute, only the inventor may apply for a patent, with certain exceptions. If a person who is not the inventor should apply for a patent, the patent, if it were obtained, would be void. The person applying in such a case would also be subject to criminal penalties for committing perjury. If the inventor is dead, the application may be made by his legal representatives, that is, the administrator or executor of his estate, in his place. If the inventor is insane, the application for patent may be made by his guardian, in his place.

If two or more persons make an invention jointly, they apply for a patent as joint inventors. A person who makes a financial contribution is not a joint inventor and cannot be joined in the application as an inventor. It is possible to correct an innocent mistake in omitting a joint inventor or in erroneously joining a person as inventor.

APPLICATION FOR PATENT

The application for patent is made to the Commissioner of Patents and includes:

(1) A written document which comprises a petition, a specification (description and claims), and an oath;

(2) A drawing in those cases in which a drawing is possible; and

(3) The Government filing fee of $30 (plus an additional $1 for each claim in excess of 20).

The petition, specification, and oath must be in the English language and must be legibly written in permanent black ink on one side of the paper. The Office prefers typewriting on legal size paper, 8 to 8½ by 12½ to 13 inches, double spaced, with margins of 1½ inches on the left-hand side and at the top. If the papers filed are not correctly, legibly, and clearly written, the Patent Office may require typewritten or printed papers.

The application for patent is not accepted and placed upon the files for examination until all its required parts, complying with the rules relating thereto, are received, except that certain minor informalities are waived subject to correction when required.

If the papers and parts are incomplete, or so defective that they cannot be accepted as a complete application for examination, the applicant will be notified; the papers will be held for six months for completion and, if not by then completed, will thereafter be returned or otherwise disposed

Guide for Patent Draftsmen

A few typical pages from *Guide for Patent Draftsmen* make up Appendix B. Each patent application must contain whatever drawings are required to explain the nature and operation of the invention. There are certain rules governing the preparation of the drawings; these are explained in the *Guide for Patent Draftsmen*. Standard electrical and mechanical symbols are shown in addition to methods of shading and perspective. How references to the drawings are made in the text of the patent is discussed. Familiarity with this pamphlet will assist the inventor in preparing the drawings that are a necessary part of the patent application.

Guide for Patent Draftsmen may be purchased for 15 cents from:

> Superintendent of Documents
> U. S. Government Printing Office
> Washington, D.C. 20402

SELECTED RULES OF PRACTICE RELATING TO PATENT DRAWINGS

THE DRAWINGS

81. Drawings required. The applicant for patent is required by statute to furnish a drawing of his invention whenever the nature of the case admits of it; this drawing must be filed with the application. Illustrations facilitating an understanding of the invention (for example flow sheets in cases of processes, and diagrammatic views) may also be furnished in the same manner as drawings, and may be required by the Office when considered necessary or desirable.

53 U. S. C. 113. Drawings. When the nature of the case admits, the applicant shall furnish a drawing.

82. Signature to drawing. Signatures are not required on the drawing if it accompanies and is referred to in the other papers of the application, otherwise the drawing must be signed. The drawing may be signed by the applicant in person or have the name of the applicant placed thereon followed by the signature of the attorney or agent as such.

83. Content of drawing. The drawing must show every feature of the invention specified in the claims. When the invention consists of an improvement on an old machine the drawing must when possible exhibit, in one or more views, the improved portion itself, disconnected from the old structure, and also in another view, so much only of the old structure as will suffice to show the connection of the invention therewith.

84. Standards for drawings. The complete drawing is printed and published when the patent issues, and a copy is attached to the patent. This work is done by the photolithographic process, the sheets of drawings being reduced about one-third in size. In addition, a reduction of a selected portion of the drawings of each application is published in the Official Gazette. It is therefore necessary for these and other reasons that the character of each drawing be brought as nearly as possible to a uniform standard of execution and excellence, suited to the requirements of the reproduction process and of the use of the drawings, to give the best results in the interests of inventors, of the Office, and of the public. The following regulations with respect to drawings are accordingly prescribed :

(a) *Paper and ink.* Drawings must be made upon pure white paper of a thickness corresponding to two-ply or three-ply Bristol board. The surface of the paper must be calendered and smooth and of a quality which will permit erasure and correction. India ink alone must be used for pen drawings to secure perfectly black solid lines. The use of white pigment to cover lines is not acceptable.

(b) *Size of sheet and margins.* The size of a sheet on which a drawing is made must be exactly 10 by 15 inches. One inch from its edges a single marginal line is to be drawn, leaving the "sight" precisely 8 by 13 inches. Within this margin all work must be included.

One of the shorter sides of the sheet is regarded as its top, and, measuring down from the marginal line, a space of not less than 1¼ inches is to be left blank for the heading of title, name, number, and date, which will be applied subsequently by the Office in a uniform style.

(c) *Character of lines.* All drawings must be made with drafting instruments or by photolithographic process which will give them satisfactory reproduction characteristics. Every line and letter (signatures included) must be absolutely black. This direction applies to all lines however fine, to shading, and to lines representing cut surfaces in sectional views. All lines must be clean, sharp, and solid, and fine or crowded lines should be avoided. Solid black should not be used for sectional or surface shading. Freehand work should be avoided wherever it is possible to do so.

(d) *Hatching and shading.* Hatching should be made by oblique parallel lines, which may be not less than about one-twentieth inch apart.

Heavy lines on the shade side of objects should be used except where they tend to thicken the work and obscure reference characters. The light should come from the upper left-hand corner at an angle of 45°. Surface delineations should be shown by proper shading, which should be open.

(e) *Scale.* The scale to which a drawing is made ought to be large enough to show the mechanism without crowding when the drawing is reduced in reproduction, and views of portions of the mechanism on a larger scale should be used when necessary to show details clearly; two or more sheets should be used if one does not give sufficient room to accomplish this end, but the number of sheets should not be more than is necessary.

(f) *Reference characters.* The different views should be consecutively numbered figures. Reference numerals (and letters, but numerals are preferred) must be plain, legible and carefully formed, and not be encircled. They should, if possible, measure at least one-eighth of an inch in height so that they may bear reduction to one twenty-fourth of an inch ; and they may be slightly larger when there is sufficient room. They must not be so placed in the close and complex parts of the drawing as to interfere with a thorough comprehension of the same, and therefore should rarely cross or mingle with the lines. When necessarily grouped around a certain part, they should be placed at a little distance, at the closest point where there is available space, and connected by lines with the parts to which they refer. They should not be placed upon hatched or shaded surfaces but when necessary, a blank space may be left in the hatching or shading where the character occurs so that it shall appear perfectly distinct and separate from the work. The same part of an invention appearing in more than one view of the drawing must always be designated by the same character, and the same character must never be used to designate different parts.

143

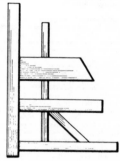

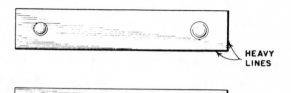

HEAVY
LINES

ALWAYS USE PLAIN BLOCK LETTERING
FOR LEGENDS NAMES, ETC.

SOME STYLES OF LETTERING
USED ON PATENT DRAWINGS

WATER
INSULATION   1234567890
COPPER
OIL

ALL FIGS. MUST BE SEPARATELY NUMBERED

Fig.1.

FIG. 1.

Fig. 2.

Fig. 2

THE LIGHT COMES
FROM THE UPPER
LEFT-HAND CORNER
AT AN ANGLE
OF 45°

ALWAYS MAKE
SHADE LINES
ON SHADOW SIDE

Letters and figures of reference must be carefully formed. Several types of lettering and figure
marks are shown, however, the draftsman may use any style of lettering that he may choose.

Place heavy lines on the shade side of objects, assuming that the light is coming from the upper
left-hand corner at an angle of 45°. Make these heavy lines the same weight throughout the
various views on the drawing.

Descriptive matter is not permitted on patent drawings. Legends may be applied when necessary
but only plain black lettering should be used.

The different views should be consecutively numbered.

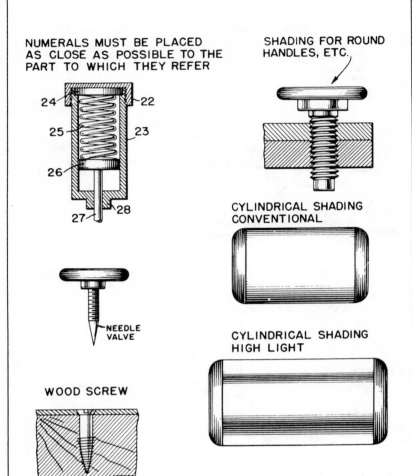

NUMERALS MUST BE PLACED
AS CLOSE AS POSSIBLE TO THE
PART TO WHICH THEY REFER

24 22
25 23
26
27 28

NEEDLE
VALVE

WOOD SCREW

SHADING FOR ROUND
HANDLES, ETC.

CYLINDRICAL SHADING
CONVENTIONAL

CYLINDRICAL SHADING
HIGH LIGHT

Reference characters should be placed at a little distance from the parts to which they refer. They should be connected with these parts by a short lead line, never by a long lead line. When necessary blank spaces must be left on shaded and hatched areas for applying the numerals.

Use wood graining sparingly on parts of wood in section. Excessive wood graining is objectionable as it blurs the view and is very confusing.

Various methods of shading are shown, however, the conventional surface shading should be used until the draftsman has obtained enough experience to attempt the more involved types of shading.

BEVEL GEARS

NOTE—TEETH OF
EACH GEAR
HAVE THE
SAME
SLANT

BALL BEARING

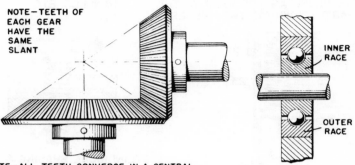

INNER
RACE

OUTER
RACE

NOTE—ALL TEETH CONVERGE IN A CENTRAL
POINT.—BROKEN LINES ARE FOR INSTRUCTION
PURPOSES AND ARE NOT TO BE PLACED ON DRAWINGS

TOP PLAN VIEW

ROLLER BEARING

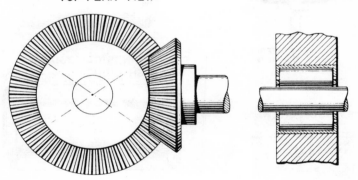

The conventional method of illustrating bevel gears is clearly shown on the two figures on the left-hand side of the page. Particular care must be given to the correct spacing between the gear teeth and also to the weight of the shade lines used. Both must be correctly shown to obtain the desired effect.

Two types of bearings are also shown. The roller bearing is clearly disclosed by the use of the conventional cylindrical shading. The fanciful black shading shown on the ball bearing is very effective in bringing out the idea of an object being shiny as well as round.

The use of white pigment to cover line is not acceptable.

SYMBOLS FOR DRAFTSMEN

Rule 84 (g) states that graphical symbols for conventional elements may be used on the drawing when appropriate, subject to approval by the Office. The symbols and other conventional devices which follow have been and are approved for such use. This collection does not purport to be exhaustive, other standard and commonly used symbols will also be acceptable provided they are clearly understood, are adequately identified in the specification as filed, and do not create confusion with other symbols used in patent drawings.

NOTES: In general, in lieu of a symbol, a conventional element, combination or circuit may be shown by an appropriately labeled rectangle, square, or circle; abbreviations should not be used unless their meaning is evident and not confusing with the abbreviations used in the suggested symbols. In the electrical symbols an arrow through an element indicates variability thereof, see for example symbols 2, 6, 12; dotted line connection of arrows indicates ganging thereof, see symbol 6; inherent property (as resistance) may be indicated by showing symbol (for resistor) in dotted lines.

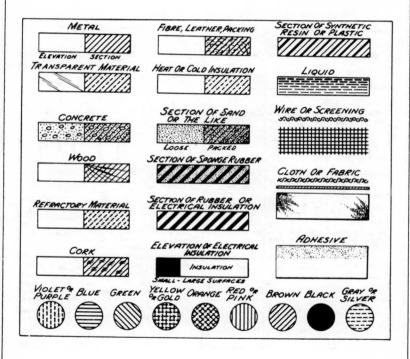

147

Electrical Symbols

RESISTOR	VARIABLE RESISTOR	POTENTIOMETER	RHEOSTATS	CONDENSERS	GANGED VARIABLE CONDENSERS
1	2	3	4	5	6
INDUCTORS	INDUCTOR ADJUSTABLE CORE	INDUCTOR OR REACTOR POWDERED MAGNETIC CORE	TRANSFORMER SATURABLE CORE	TRANSFORMER AIR CORE	VARIABLE TRANSFORMER
7	8	9	10	11	12
TRANSFORMER MAGNETIC CORE	AUTO-TRANSFORMER ADJUSTABLE	CROSSED AND JOINED WIRES	MAIN CIRCUITS / SHUNT OR CONTROL CIRCUITS	FUSE	COAXIAL CABLES
13	14	15	16	17	18
SHIELDING	BATTERY	THERMOELEMENT	BELL	AMMETER	MILLIAMMETER
19	20	21	22	23	24
VOLTMETER	GALVANOMETER	WATTMETER	SWITCH	DOUBLE POLE SWITCH	DOUBLE POLE DOUBLE THROW SWITCH
25	26	27	28	29	30
PUSH BUTTON TWO POINT MAKE	SELECTOR OR CONNECTOR OR FINDER SWITCH	CIRCUIT BREAKER OVERLOAD	RELAY	POLARIZED RELAY	DIFFERENTIAL RELAY
31	32	33	34	35	36
ANNUNCIATORS SIDE FRONT	DROP ANNUNCIATOR	DRUM TYPE SWITCH OR CONTROL 1 2 1 2	COMMUTATOR MOTOR OR GENERATOR	REPULSION MOTOR	INDUCTION MOTOR THREE PHASE SQUIRREL CAGE
37	38	39	40	41	42
INDUCTION MOTOR PHASE WOUND SECONDARY	SYNCHRONOUS MOTOR OR GEN. THREE PHASE	MOTOR GENERATOR	ROTARY CONVERTER THREE PHASE	FREQUENCY CHANGER THREE PHASE	TROLLEYS
43	44	45	46	47	48
THIRD RAIL SHOE	RECEIVERS	TRANSMITTER OR MICROPHONE	TELEPHONE HOOK	TELEGRAPH KEY	SWITCH BOARD PLUG AND JACK
49	50	51	52	53	54

Electrical Symbols – continued

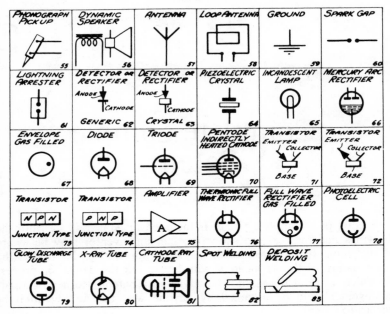

PHONOGRAPH PICK UP 55	DYNAMIC SPEAKER 56	ANTENNA 57	LOOP ANTENNA 58	GROUND 59	SPARK GAP 60
LIGHTNING ARRESTER 61	DETECTOR or RECTIFIER — ANODE — CATHODE — GENERIC 62	DETECTOR or RECTIFIER — ANODE — CATHODE — CRYSTAL 63	PIEZOELECTRIC CRYSTAL 64	INCANDESCENT LAMP 65	MERCURY ARC RECTIFIER 66
ENVELOPE GAS FILLED 67	DIODE 68	TRIODE 69	PENTODE INDIRECTLY HEATED CATHODE 70	TRANSISTOR EMITTER COLLECTOR BASE 71	TRANSISTOR EMITTER COLLECTOR BASE 72
TRANSISTOR N P N JUNCTION TYPE 73	TRANSISTOR P N P JUNCTION TYPE 74	AMPLIFIER A 75	THERMIONIC FULL WAVE RECTIFIER 76	FULL WAVE RECTIFIER GAS FILLED 77	PHOTOELECTRIC CELL 78
GLOW DISCHARGE TUBE 79	X-RAY TUBE 80	CATHODE RAY TUBE 81	SPOT WELDING 82	DEPOSIT WELDING 83	

Mechanical Symbols

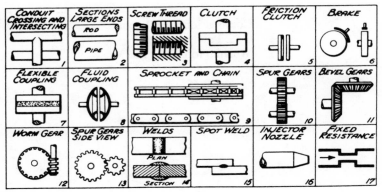

CONDUIT CROSSING AND INTERSECTING 1	SECTIONS LARGE ENDS ROD PIPE 2	SCREW THREAD 3	CLUTCH 4	FRICTION CLUTCH 5	BRAKE 6
FLEXIBLE COUPLING 7	FLUID COUPLING 8	SPROCKET AND CHAIN 9		SPUR GEARS 10	BEVEL GEARS 11
WORM GEAR 12	SPUR GEARS SIDE VIEW 13	WELDS PLAN SECTION 14	SPOT WELD 15	INJECTOR NOZZLE 16	FIXED RESISTANCE 17

Complete Sample Patent

A copy of a complete patent is included in Appendix C. Both the drawings and the text are shown so that an inventor may be aware of the detail that is required in a patent application. Naturally, the exact nature of the application will depend on the device or process on which a patent is sought. The advice of a patent attorney or a patent agent should be obtained by anyone who has an invention he wishes to patent.

Copies of most U.S. Patents and lists of patents in various classifications are available from:

> Commissioner of Patents
> U. S. Patent Office
> Washington, D.C. 20231

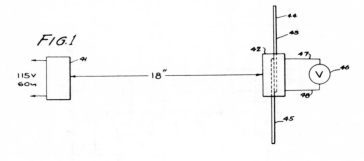

FIG.1

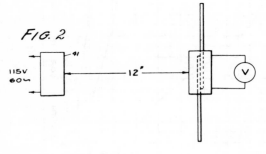

FIG. 2

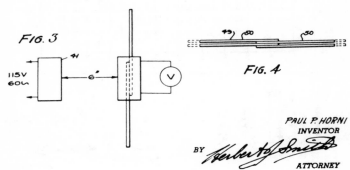

FIG. 3

FIG. 4

PAUL P. HORNI
INVENTOR

BY *Herbert J. Smith*

ATTORNEY

151

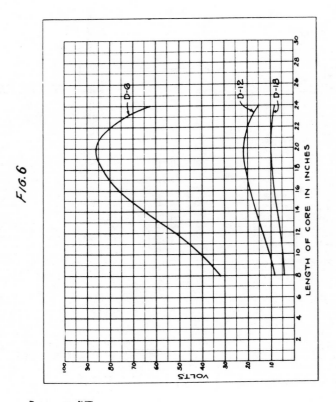

FIG. 6

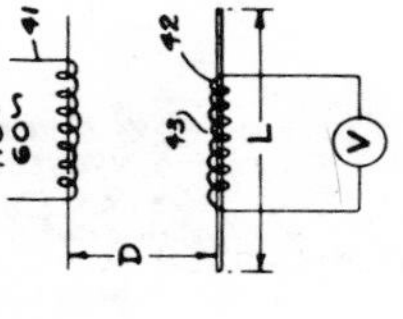

FIG. 5

PAUL P. HORNI
INVENTOR

BY

ATTORNEY

UNITED STATES PATENT OFFICE

2,378,014

METHOD OF DETERMINING THE EFFECTIVE RANGE OF MAGNETIC DETECTORS

Paul P. Horni, Newark, N. J.

Application March 24, 1942, Serial No. 435,922

4 Claims. (Cl. 29—155.5)

Fundamentally, the invention pertains to a vehicle actuable device which utilizes a magnetic detector, which detector is energized by the passage of a vehicle to operate an electro-sensitive element which in turn initiates operation of a control apparatus. The control apparatus may be of varying types such as a traffic controller for energizing traffic signals at an intersection of a street.

The device may also be used for permitting a vehicle to initiate operation of a mechanism to detonate an explosive charge for destruction and demolition of a vehicle, such as a tank. The device also may be used for demolition of bridges and other structures over which tanks and other mechanized units may seek to traverse.

Heretofore, when employing devices which are actuable by a detector of the magnetic type, various techniques have been used in fabricating the magnetic detector. One of the types of magnetic detectors consists primarily of a coil of wire wrapped about a core of magnetizable material, the core and the coil of wire having certain characteristics so that it will operate under given conditions. Ordinarily the core is of a length which has been determined by the cut and try method. The length of the core being usually sufficient to produce satisfactory operation with the particular equipment with which it is used.

The present invention contemplates a core which is variable in length so that the length of said core may be varied to produce different degrees of sensitivity, for example, with a specified type of electro-sensitive element and control apparatus the length of the core of the detector may be varied so that the apparatus will operate efficiently at a given distance, let us say 2 ft., and the length of the core may then be varied so that the same vehicle will actuate the apparatus at a distance of let us say 10 ft.

It will therefore be seen that the present invention may be used to vary the efficiency and effectiveness of a magnetic detector by varying the length of the core. This becomes extremely essential in magnetically actuable detecting equipment such as may be used for military purposes.

With certain types of magnetically actuable equipment for military purposes the detector may be so adjusted that the ordinary tank must be within 2 ft. of the detector in order to generate sufficient voltage to actuate the detector and its equipment responsive thereto. The present invention may utilize this same equipment but with a longer core so that a vehicle would detonate a mine at 10 ft. or 12 ft. or even greater distances,

provided the charge of the main explosive is sufficient to accomplish the desired demolition objective.

It is common practice in signaling systems, particularly those systems relating to the control of traffic of a vehicular type, namely, automobiles and the like, to employ what is commonly referred to as detectors. The detectors are generally classified as either mechanical or magnetic. In the mechanical type of detector, the vehicle must actually engage a contactor which is located in the street, in the path of the vehicle. As the vehicle engages the contactor, an electrical circuit is completed, which causes certain apparatus to function, which, through a series of operations, eventually causes a traffic signal to be operated. In the magnetic type of detector, a device is placed in the street in the vicinity of the path of the vehicle to be detected, said detector being actuated by the passage thereover or thereabouts, due to the presence of a mass of magnetizable material, the vehicle, per se, not physically engaging or pressing any tangible element. While it is old in the art to use a coil with a simple core as a magnetic detector, this is, perhaps, one of the simplest forms of magnetic detector.

It is further pointed out that while there have been magnetic detectors which were usable to give certain results for the control of signaling apparatus, there are many cases where the vehicle itself carried a permanent magnet for the purpose of generating a current in a sensitive instrument, due to the cutting of the magnetic field by conductors of the coil of said magnetic detector. To provide a detector which requires this type of permanent magnet installation on a vehicle, is definitely costly in itself, and further undesirable, due to the installation charges.

A great deal of experimentation and scientific research has disclosed that all vehicles are fundamentally and inherently magnetized. This may be due to many causes, such as the aligning of the parts of the metal used in construction to form resultant magnetic poles, or partly due to the formation of a number of independent magnets. It is well known that a ship will take on a given polarity, due to the position in which it rests when actual construction has taken place. From many causes a vehicle, therefore, is a huge magnet, or a multiplicity of magnets. Tests indicate that a front and rear bumper of an automobile may each be independent magnets and that the vehicle itself will have many opposing polarities, which may be manifested by moving a compass in the vicinity of various parts of the vehicle. Regardless of the peculiar polarities of various vehicles, it is nevertheless established by tests that the vehicle in itself does possess enough magnetic material to produce magnetizing lines of force emanating therefrom to cause a current to flow or a voltage to be set up in a detector of the type herein employed.

It is an object of the present invention to provide a novel method and means of increasing the efficiency of a magnetic detector and vehicle actuable detector system utilizing said detector.

It is a further object of the invention to provide a method of determining the maximum efficiency of a magnetic detector by varying the length of the core of the detector.

It is a further object of the invention to provide a novel method of determining the maximum efficiency of a magnetic detector by establishing the length of the core in relation to a given coil or range of coil lengths.

Other and further objects may be and may become apparent to one skilled in the art from a perusal of the drawings and specifications presented herewith and the present disclosure is not to be considered as a limitation since it is illustrative of only certain embodiments of the invention.

The magnetic detector and signalling system therefor disclosed in this application are claimed in my application Serial Number 478,850, filed March 11, 1943.

In the drawings

Figs. 1, 2 and 3 are schematic representations of a method of determining the efficiency of the magnetic detector as controlled by the length of the core.

Fig. 4 is an arrangement of the core laminations per se being shown dotted indicating the elongation of the overall core.

Fig. 5 is a schematic diagram of various coils electrically connected during the tests as shown in Figs. 1, 2 and 3.

Fig. 6 is a graph showing the various values of voltage generated in the detector as the length of the core and the distance between the detector and the source of energy is varied.

While various vehicles such as automobiles, tanks, armored units and other mobile devices are known to possess a multiplicity of magnetic poles, it can be generally said that while no two vehicles have the same external magnetic characteristics, a certain type vehicle will have characteristics and magnetic strength generally similar to other vehicles of the same type. In systems employing the magnetic detector as set forth herein, the system is generally arranged so that the detector in combination with other equipment will "pick up" the vehicle having the weakest magnetic strength effective to initiate operation of the system in whatever form it may be.

In order to determine the results as presented in the present application, a consistent field strength has been utilized as presented in Figs. 1, 2 and 3 by energizing a coil 41 with 115 volts of alternating current having a frequency of 60 cycles.

In Figs. 1, 2 and 3 the coil 41 having a substantially constant field strength is representative of a unit of traffic. In Fig. 1 the detector 42 having a core 43 with laminations 44 and 45 being contiguous and overlapping within the confines of the core. A voltmeter 46 is serially connected to the winding of the detector 42, by conductors 47 and 48. With the distance shown

as 18", the voltage generated in the detector 42 was measured. The voltages in the detector 42 were measured with varying overall lengths of the core 43.

In Fig. 2 the same equipment was used but the spacing between the coil 41 and the detector 42 was reduced to 12" and the tests again repeated in the same manner and those carried out for the disclosure in Fig. 1.

Fig. 3 shows a representation where the distance between the coil 41 and the detector 42 was reduced to 6" and the tests again repeated.

In all of the tests the measurement between the coil 41 and the detector 42 was from the axial center of the coil 41 to the axial center of the detector 42. The distances of 6", 12" and 18" for the respective views shown in Figs. 3, 2 and 1 respectively are distances from the axial center of the coil 41 to the axial center of detector 42 in each instance. The results of the tests were all plotted on a chart as shown in Fig. 6.

From observing the chart it will be seen that the curves D—18, D—12 and D—6 were all made with cores ranging from 8" in length to 24" in length, and in each instance laminated cores were used as well as a single core for each of the tests.

By observing the chart in Fig. 6, it will be seen that as the core length was increased from the minimum of 8" to the maximum of 24", the greatest voltage induced in the detector was attained with a core of approximately 20" in length. After the core length was extended beyond the 20" length, the efficiency decreased. Cores over 24" in length were not used as it was obvious that the efficiency decreased approximately the same amount as increased.

Curve D—6 being farthest away from the coil 41, had only approximately an increase in 12 volts from the 8" core to the maximum efficiency of the 20" core. The increase in voltage generated in the detector when the distance of 12" is shown on the curve D—12 of about 14 volts when the core ranges from 8" to 20", while the D—6 curve shows a variation in generated voltage of about 54 volts.

The coil 41 and the coil of the detector 42 were varied in size throughout the various tests. Coils in both instances were varied from 3" to about 8" in length and the various length coils 41 were interchanged with various length coils of detector 42 over the same range of coil length.

In all of the tests there is a definite indication that the size of the coil used in ordinary detectors of the type set forth herein did not have any appreciable effect upon the results obtained concerning the core length, so that it might be generally set forth that the shape and size of the coil has no apparent effect on this phenomena.

It is evident from the graph that the square law holds true showing that as the distance is doubled you receive one-fourth of the voltage. For instance, if we use a core length of 20", we have a voltage of 85 volts when the detector is 6" from the source and we have a voltage of 21½ volts when the detector is 12" from the source.

As far as can be determined, the length of the coil has no bearing on the phenomenon of a definite maximum length of core as indicated in the graph as approximately 20"

Fig. 4 shows a core 49 composed of individual laminations 50 which may be extended as shown by the dotted lines at the ends of the core. This represents generally that the various laminations are placed in touching relation with each other and are overlapping so that while the core laminations themselves are separate, the effect is the same as though the core were a continuous piece of material. In normal operation the core laminations may be bound together by tape or may be held by any mechanical device which is convenient or suitable to be disposed within the coil of the detector.

In Fig. 5 a schematic representation of the coil 41 is shown with the detector 42 having the winding 43 thereon with the legend L as representative of the overall core length, while the legend D is representative of the distance between the centers of the coil 41 and the detector 42. Fig. 5 shows generally the circuit and the method of determining the results obtained, as shown by the three curves on the chart in Fig. 6.

From the foregoing it will be seen that a detector may be used to vary the efficiency of a control system, such as a traffic control, by varying the length of a detector, which may be assumed as the length of the coil of the detector, to a maximum of approximately 20". In operation, when used in a traffic signal, a core of a given length may be used to control a single lane of traffic which may be assumed as being approximately 9' wide, and a vehicle in an adjacent lane will not actuate the traffic signal responsive to said detector.

Since there are many variables in a system of this type it can be said generally that all magnetic detectors which depend on a core for suitable operation may have the core length limited to approximately 20" in length for maximum efficiency. By using a shorter core length the effective range will be decreased and this core length must be determined for the particular job at hand. If it is for traffic control, the core length may be set to operate over one, two or three lanes of traffic, but if it is for military use, the core length would be determined by the approximate distance the vehicle must be from the detector before a mine will explode, this distance being generally determined by a study of the largest tank or vehicle which is to be demolished or immobilized by the discharge of the mine.

What I claim as new and desire to secure by Letters Patent of the United States is:

1. A method of determining the desired effective range of a magnetic detector by disposing a source of magnetic energy at a given distance from the coil of a magnetic detector; inserting overlapping laminations within the coil of the detector to form a core for said detector, and moving the overlapping core laminations inwardly or outwardly until the desired voltage is generated in the magnetic detector dependent upon the overall length of the core, and securely binding the laminations together within the core, whereby the desired effective range of the detector is known.

2. A method of determining the desired effective range of a magnetic detector by disposing a source of magnetic energy at a given distance from the coil of a magnetic detector, inserting overlapping laminations within the coil of the detector to form a core for said detector, and moving the core laminations inwardly or outwardly until the desired voltage is generated in the magnetic detector dependent upon the overall length of the core, securely binding the laminations together within the core, and securely positioning the core at approximately the center of the magnetic detector coil to hold said core in fixed relation to the detector coil, to the end that the desired effective range of the detector is known.

3. A method of determining the desired effective range of a magnetic detector by disposing a coil at a given distance from the coil of a magnetic detector, energizing said first coil with alternating current, inserting overlapping laminations within the coil of the detector to form a substantially unitary core for said detector, and moving the core inwardly or outwardly until the desired voltage is generated in the magnetic detector dependent upon the length of the core, and securing the laminations together within the core to maintain the desired generated voltage, whereby the effective range of the magnetic detector is known.

4. A method of determining the desired effective range of a magnetic detector by disposing a coil at a given distance from the coil of a magnetic detector, energizing said first coil with a source of alternating current, inserting overlapping laminations within the coil of the detector to form a core for said detector, and moving the core inwardly or outwardly until the desired voltage is generated in the magnetic detector dependent upon the length of the core, securely binding the laminations together within the core, and securely positioning the core at approximately the center of the magnetic detector coil to maintain the desired generated voltage to the end that the effective range of the magnetic detector is known.

PAUL P. HORNI.

General Information Concerning Trademarks

The opening remarks from *General Information Concerning Trademarks* are contained in Appendix D. The publication is of interest to those who have elected to manufacture and sell their inventions directly to consumers. Trademarks identify products from competitive makes and promote repeat orders.

General Information Concerning Trademarks may be purchased for 35 cents from:

Superintendent of Documents
U. S. Government Printing Office
Washington, D.C. 20402

Trademark
Statutes and Rules

This pamphlet is intended to serve only as a general guide in regard to trademark matters in the Patent Office.

Applications for registration of trademarks must conform to the requirements of the Trademark Act of 1946, as amended, and the Trademark Rules of Practice. This Act, Public Law 489, Seventy-ninth Congress, Chapter 540, 60 Stat. 427, popularly known as the Lanham Act, and forming Chapter 22, Title 15 of the U.S. Code, became effective July 5, 1947, superseding the Trademark Acts of 1905 and 1920. A copy of the *Trademark Rules of Practice With Forms and Statutes* can be obtained from the Superintendent of Documents, Washington, D.C., 20402, for 45¢.

Definition and Functions of Trademarks

Definition of Trademarks. A "trademark," as defined in section 45 of the 1946 Act, "includes any word, name, symbol, or device, or any combination thereof adopted and used by a manufacturer or merchant to identify his goods and distinguish them from those manufactured or sold by others."

Function of Trademarks. The primary function of a trademark is to indicate origin. However, trademarks also serve to guarantee the quality of the goods bearing the mark and, through advertising, serve to create and maintain a demand for the product. Rights in a trademark are acquired only by use and the use must ordinarily continue if the rights so acquired are to be preserved. Registration of a trademark in the Patent Office does not in itself create or establish any exclusive rights, but is recognition by the Government of the right of the owner to use the mark in commerce to distinguish his goods from those of others.

Mark Must Be Used in Commerce. In order to be eligible for registration, a mark must be in use in commerce which may lawfully be regulated by Congress, for example, interstate commerce, at the time the application is filed. "Use in commerce" is defined in section 45 as follows:

For the purposes of this Act a mark shall be deemed to be used in commerce (a) on goods when it is placed in any manner on the goods or their containers or the displays associated therewith or on the tags or labels affixed thereto and

Disclosure Document Program

Appendix E consists of an abstract of a new service to inventors, the U.S. Patent Office *Disclosure Document Program*. This service provides acceptance and preservation for a period of two years of evidence of the dates of conception of inventions. While submissions may be made in a relatively informal manner, the actual application for patent must be filed within the two-year period. A filing fee of $10 is charged.

The pamphlet *Disclosure Documents Program* may be obtained at a cost of 15 cents from:

> Superintendent of Documents
> U.S. Government Printing Office
> Washington, D.C. 20402

A new service is provided for inventors by the U. S. Patent Office—the acceptance and preservation for a limited time of "Disclosure Documents" as evidence of the dates of conception of inventions.

WHAT THE PROGRAM IS

A paper disclosing an invention and signed by the inventor or inventors may be forwarded to the Patent Office by the inventor (or by any one of the inventors when there are joint inventors), by the owner of the invention, or by the attorney or agent of the inventor(s) or owner. It will be retained for two years and then be destroyed unless it is referred to in a related patent application filed within two years.

The Disclosure Document is not a patent application, and the date of its receipt in the Patent Office will not become the effective filing date of any patent application subsequently filed. However, like patent applications, these documents will be kept in confidence by the Patent Office.

This program does not diminish the value of the conventional witnessed and notarized records as evidence of conception of an invention, but it should provide a more creditable form of evidence than that provided by the popular practice of mailing a disclosure to oneself or another person by registered mail.

CONTENT OF DISCLOSURE DOCUMENT

Although there are no restrictions as to content and claims are not necessary, the benefits afforded by the Disclosure Document will depend directly upon the adequacy of the disclosure. Therefore, it is strongly urged that the document contain a clear and complete explanation of the manner and process of making and using the invention in sufficient detail to enable a person having ordinary knowledge in the field of the invention to make and use the invention. When the nature of the invention permits, a drawing or sketch should be included. The use or utility of the invention should be described, especially in chemical inventions.

PREPARATION OF THE DOCUMENT

The Disclosure Document must be limited to written matter or drawings on paper or other thin, flexible material, such as linen or plastic drafting material, having dimensions or being folded to dimensions not to exceed 8½ by 13 inches. Photographs also are acceptable. Each page should be numbered. Text and drawings should be sufficiently dark to permit reproduction with commonly used office copying machines.

OTHER ENCLOSURES

In addition to the fee described below, the Disclosure Document *must* be accompanied by a stamped, self-addressed envelope and a separate paper in duplicate, signed by the inventor, stating that he is the inventor and requesting that the material be received for processing under the Disclosure Document Program. The papers will be stamped by the Patent Office with an identifying number and date of receipt, and the duplicate request will be returned in the self-addressed envelope together with a notice indicating that the Disclosure Document may be relied upon only as evidence and that a patent application should be diligently filed if patent protection is desired. The inventor's request may take the following form:

"The undersigned, being the inventor of the disclosed invention, requests that the enclosed papers be accepted under the Disclosure Document Program, and that they be preserved for a period of two years."

DISPOSITION

The Disclosure Document will be preserved in the Patent Office for two years and then will be destroyed unless it is referred to in a related patent application filed within the two-year period. The Disclosure Document may be referred to by way of a letter of transmittal in a new patent application or by a separate letter filed in a pending application. Unless it is desired to have the Patent Office retain the Disclosure Document beyond the two-year period, it is not required that it be referred to in a patent application.

Literature References For Inventors

Literature references regarding inventions and inventing make up Appendix F. The books listed pertain to the recent history of invention, methods of approach, overall background research, sales aspects, and the like. The perusal of this literature is recommended for every earnest inventor desiring to save time, labor, and money.

GENERAL LITERATURE FOR INVENTORS

V. D. Angerman, *How to Find a Buyer for Your Invention* (New York: Science and Mechanics Publishing Co., 1956).

R. Burlingame, *Scientists Behind the Inventors* (New York: Avon Books, 1965).

J. A. Kuecken, *Creativity, Invention, & Progress* (Indianapolis: Howard W. Sams & Co., Inc., 1969).

J. Rossman, *Industrial Creativity: The Psychology of the Inventor* (New Hyde Park, New York: University Books, Inc., 1964).

J. W. N. Sullivan, *The Limitations of Science* (New York: New American Library, Inc., 1963).

E. B. Wilson, *An Introduction to Scientific Research* (New York: McGraw-Hill Book Company, 1952).

M. Wilson, *American Science and Invention* (New York: Simon & Schuster, Inc., 1954).

R. F. Yates, *Yates' Guide to Successful Inventing* (New York: Funk & Wagnalls Co., 1967).